SpringerBriefs in Environment, Security, Development and Peace

Mediterranean Studies

AF801011

Volume 7

Series Editor

Hans Günter Brauch

For further volumes:
http://www.springer.com/series/11792
http://www.afes-press-books.de/html/SpringerBriefs_ESDP.htm

Ebru A. Gencer

The Interplay Between Urban Development, Vulnerability, and Risk Management

A Case Study of the Istanbul Metropolitan Area

Springer

Ebru A. Gencer
CUDRR
New York
USA

ISSN 2196-5469 ISSN 2196-5447 (electronic)
ISBN 978-3-642-29469-3 ISBN 978-3-642-29470-9 (eBook)
DOI 10.1007/978-3-642-29470-9
Springer Heidelberg New York Dordrecht London

Library of Congress Control Number: 2013936953

© The Author(s) 2013
This work is subject to copyright. All rights are reserved by the Publisher, whether the whole or part of the material is concerned, specifically the rights of translation, reprinting, reuse of illustrations, recitation, broadcasting, reproduction on microfilms or in any other physical way, and transmission or information storage and retrieval, electronic adaptation, computer software, or by similar or dissimilar methodology now known or hereafter developed. Exempted from this legal reservation are brief excerpts in connection with reviews or scholarly analysis or material supplied specifically for the purpose of being entered and executed on a computer system, for exclusive use by the purchaser of the work. Duplication of this publication or parts thereof is permitted only under the provisions of the Copyright Law of the Publisher's location, in its current version, and permission for use must always be obtained from Springer. Permissions for use may be obtained through RightsLink at the Copyright Clearance Center. Violations are liable to prosecution under the respective Copyright Law.
The use of general descriptive names, registered names, trademarks, service marks, etc. in this publication does not imply, even in the absence of a specific statement, that such names are exempt from the relevant protective laws and regulations and therefore free for general use.
While the advice and information in this book are believed to be true and accurate at the date of publication, neither the authors nor the editors nor the publisher can accept any legal responsibility for any errors or omissions that may be made. The publisher makes no warranty, express or implied, with respect to the material contained herein.

Printed on acid-free paper

Springer is part of Springer Science+Business Media (www.springer.com)

Acknowledgments

The research in this book was primarily accomplished for my doctoral dissertation (Gencer 2007) at Columbia University and it has been possible with the help and support of many individuals and institutions. First, I must thank my dissertation committee and most importantly my sponsor, Elliott Sclar, who supported me throughout my tenure at Columbia University's Urban Planning Program.

ProVention Consortium provided me funding to conduct field research in Turkey. I would like to thank the Consortium and the World Bank, which facilitated the development of my research through their Web-based discussions and forums in the Young Professionals Applied Grants Program.

My research in Istanbul would not have been possible without the help and assistance of my interview respondents in district municipalities. In Istanbul, I would also like to acknowledge the support of my colleagues at Mimar Sinan University's Urban Planning Program, and of Gülay Barbarosoĝlu at Boĝaziçi University's CENDIM (Center for Disaster Management) and Mustafa Erdik at its Kandilli Observatory.

I am also thankful to the editor of the Springer Series, Hans Gunter Brauch, who gave me the opportunity to publish this research on the Springer Briefs on ESDP. I should acknowledge that previous versions of parts of my research in this book are published in Gencer (2008, 2011).

I illustrated this book with my own figures and graphs (Figs. 2.1, 2.2, 2.3, 3.15, 4.2), graphical adaptation (Figs. 3.9, 3.12, 4.3), and photographs (Figs. 2.4, 2.5, 2.6, 2.7, 3.8, 3.13, 3.14, 4.4, 4.5, 4.6). Several figures are in the public domain (Figs. 3.7, 3.9, 3.10, 3.11, 3.16, 3.17, 4.1, 4.7) or used with the courtesy of the organizations which hold copyrights (Figs. 3.1, 3.2, 3.3). One figure (Fig. 3.5) is used with the courtesy of and as a registered member of the Earthquake Engineering Research Center of the University of California, Berkeley. For the remaining (Figs. 3.4 and 3.6), I obtained permission from the copyright holder.

The aim of this book is to make a difference in the way our cities can be planned and managed, reducing disaster risks and creating sustainable communities. I would like to dedicate this book to a disaster-free future for Istanbul and Istanbulites.

References

Gencer EA (2007) The interplay between natural disasters, vulnerability, and sustainable development. Ph.D diss. New York, Columbia University

Gencer EA (2008) Natural disasters, vulnerability, and sustainable development: examining the interplay, global trends, and local practices in Istanbul. VDM Verlag, Saarbrucken

Gencer EA (2011) Risk management strategies for the predicted earthquake hazard in Istanbul. In: Brauch HG, Spring OU, Kameri-Mbote P, Mesjasz C, Grin J, Chourou B, Dunay P, Birkmann J (eds) Coping with global environmental change, disasters and security threats, challenges, vulnerabilities and risks. Springer, New York, pp 655–667

Contents

Abbreviations

AFAD	Province Disaster and Emergency Directorate [Il Afet ve Acil Durum Mudurlugu]
AICP	American Institute of Certified Planners
AKOM	Istanbul Metropolitan Municipality Disaster Coordination Centre [Afet Koordinasyon Merkezi]
APA	American Planning Association
AUDMP	Asian Urban Disaster Mitigation Program
AVSI	Association of Volunteers for International Service
AYM	Istanbul Governorship Disaster Management Centre [Afet Yönetim Merkezi]
BIMTAS	Istanbul Engineering and Consultancy Services Cooperation [Boğaziçi Insaat ve Muhendislik Anonim Sirketi]
BU	Boğaziçi University [Boğaziçi Üniversitesi]
CENDIM	Center for Disaster Management
CONDER	Urban Development Company of the State of Bahai
CRED	Center for Research on Epidemiology of Disasters
CU	Columbia University
DASK	Natural Disasters Insurance Institution [Dogal Afet Sigortalari Kurumu]
DMMP	Disaster Management Master Plan
DPT	Turkey's State Planning Organization [Devlet Planlama Teşkilatı]
EERI	Earthquake Engineering Research Institute
EM-DAT	Emergency Database
EMI	Earthquakes and Megacities Initiative
EMPI	Earthquake Mitigation Plan for Istanbul
EMS	European Macro-Seismic Scale
FEMA	Federal Emergency Management Agency
GDP	Gross Domestic Product
GFDRR	Global Facility for Disaster Risk Reduction
GIS	Geographical Information Systems

HAZUS	Hazard US
İB	Istanbul Municipality [Istanbul Belediyesi]
İBB	Istanbul Metropolitan Municipality [Istanbul Büyükşehir Belediyesi]
IBHS	Institute for Business and Home Safety
ICLEI	International Council for Local Environmental Initiatives
IEMP	Istanbul Earthquake Master Plan
IMM	Istanbul Metropolitan Municipality
IMP	Istanbul Metropolitan Planning and Design Center [Istanbul Metropolitan Planlama ve Tasarim Merkezi]
IPCC	Intergovernmental Panel on Climate Change
ISOCARP	International Society of City and Regional Planning
İTÜ	Istanbul Technical University [İstanbul Teknik Üniversitesi]
JICA	Japan International Cooperation Agency
KHK	Decree Law [Kanun Hukmunde Kararname]
METU	Middle East Technical University [Orta Doĝu Teknik Üniversitesi]
NASA	National Aeronautics and Space Administration
NEHRP	National Earthquake Hazards Reduction Program
NGO	Non-Governmental Organization
NIBS	National Institute for Building Sciences
NSF	National Science Foundation
OECD	Organisation for Economic Co-operation and Development
OFDA	Office of Foreign Disaster Assistance
OYO	OYO Corporation
PCI	Pacific Consultants International
SPARC	Society for the Promotion of Area Resource Centres
SPDMI	Strategic Plan for Disaster Mitigation in Istanbul
TAY	Turkish Emergency Management Directorate [Türkiye Afet Yönetimi]
TCBİB	Turkish Republic Ministry of Development [Turkiye Cumhuriyeti Bayındırlık ve İskan Bakanlıĝı]
TCİBB	Turkish Republic Istanbul Metropolitan Municipality [Turkiye Cumhuriyeti İstanbul Büyükşehir Belediyesi]
TCIP	Turkish Catastrophe Insurance Pool
TCİV	Turkish Republic Istanbul Governorate [Turkiye Cumhuriyeti İstanbul Valiliĝi]
TMMOB ŞPO	Turkish Union of Architects and Engineers, Chamber of Planners [Türkiye Mimarlar Muhendisler Odalar Birligi, Şehir Plancıları Odası]
TPOB	Turkish Union of Planning Schools [Türkiye Planlama Okulları Birliĝi]
TRGIM	Turkish Republic Greater Istanbul Municipality [Türkiye Cumhuriyeti Büyük Istanbul Belediyesi]

TUSIAD	Turkish Industrialist's and Businessmen's Association [Türk Sanayicileri ve İşadamaları Derneği]
UN	United Nations
UNDP	United Nations Development Programme
UN-HABITAT	United Nations Human Settlements Programme
UNISDR	United Nations International Strategy for Disaster Reduction
UNPD	United Nations Population Division
US	United States (of America)
USGS	United States Geological Survey
WB	World Bank
WCED	World Commission on Environment and Development
YTÜ	Yildiz Technical University [Yıldız Teknik Üniversitesi]

About the Book

Good urban planning and risk management practices can be a powerful catalyst for reducing losses from natural disasters while simultaneously helping to develop a sustainable environment. This book illustrates the crucial interplay between urban development, risk management, and vulnerability from natural disasters with a theoretical overview and an in-depth case study in the earthquake prone city of Istanbul. The theoretical overview reviews the socio-economic, spatial, and institutional factors that create disaster vulnerability in urban areas and examines various policy and programs in play for disaster risk reduction. The empirical case study in Istanbul begins by examining the role of urban development in creating the current socio-economic and spatial vulnerability in the city, and then continues with a study of urban and risk management activities following the 1999 Marmara Earthquakes in Turkey. The book concludes that the success of urban planning and risk management actions and policies not only lies in providing sustainable solutions to urban dynamics, but also in their execution with good governance in urban areas.

Chapter 1
Introduction

Natural disasters[1] are increasingly affecting the world, taking lives unexpectedly, and causing many other injured and homeless people. They disrupt local, national and even global economies, with the capacity to change the direction of development. Statistics indicate a rising trend in the number of disasters and its impacts—affected population and monetary damage—especially within the last two decades[2] (See WB 2010 and UN 2011). In 2011 alone, natural disasters affected 98 countries, killing 30 thousand and affecting over 200 million people, and resulting in record 366 billion US dollars in economic damages (CRED 2012: 1). Adding to the current situation, the *Special Report of the Intergovernmental Panel on Climate Change* models project "substantial warming in temperature extremes," "likely" increase in "the frequency of heavy precipitation" and "tropical cyclone wind speed," and "upward trends in extreme coastal high water," all pointing out to the increase in the occurrence and severity of climatological and weather related hazards in the 21st century (IPCC 2012: 9–13).

Urban settlements are particularly vulnerable from the effects of natural hazards. Concentration of substandard infrastructure and housing, material assets, and inherent socio–economic inequalities increase susceptibility in large urban areas. Natural disasters contribute to further social, physical, and economic impediments in the sustainable development[3] of urban settlements.

---

[1] In this book, the term *natural* is used to represent the type of hazard that inflicts a disaster; it does not imply that disasters occur naturally. These terms will be discussed in detail in the second chapter.

[2] This is suggested to be related to "greater exposure, more reporting, or a combination of both" (WB 2010: 26–27).

[3] The most commonly cited definition of *sustainable development* is, "development which meets the needs of the present without comprising the ability of future generations to meet their needs" (WCED 1987: 43). This book uses the term as the integrity of an equitable and long-term economic, social, and physical development that is in accordance with the environment.

E. A. Gencer, *The Interplay Between Urban Development, Vulnerability, and Risk Management*, Mediterranean Studies 7,
DOI: 10.1007/978-3-642-29470-9_1, © The Author(s) 2013

Good urban planning[4] can be a powerful catalyst for reducing losses from natural disasters, while simultaneously helping to develop a sustainable environment. Yet, the existing situation indicates that sustainable urban development and risk management measures are not taken into consideration or may not be put into practice for a variety of financial, political, and social reasons. The interplay between urban development, vulnerability, and risk management presents itself here. With a theoretical overview and two case studies in the metropolitan city of Istanbul, this book shows that socio–economic, spatial, and institutional disparities can increase vulnerability and risk in hazard-prone urban areas, putting a setback on successful risk management and sustainable urban development.

1.1 Research Questions and Methods

This book is based on the author's doctoral dissertation, which examined the interplay between natural disasters, vulnerability, and sustainable development (Gencer 2007). As part of a larger framework, this book builds on to investigate how the sustainability of an urban environment shapes its vulnerability from natural disasters. This book is also concerned with understanding the extent of impact natural disasters have on disaster risk management, as well as investigating factors that influence the adoption and the implementation of disaster risk reduction strategies in urban areas.

The book undertakes a qualitative research which examines how urban development and management can shape the way cities are affected by natural hazards. The study indicates a significant interrelation between sustainable urban development and vulnerability from natural disasters. It demonstrates that inequitable and one-dimensional planning and development policies can increase disaster vulnerability and adversely influence the sustainability of a region. It also presents that socio–economic disparities create differences in the way local governments and urban communities adopt and employ risk and urban management activities.

A number of disaster research theorists have stated the link between sustainable urban development and natural disasters. No research, however, to the author's knowledge, has yet been accomplished by way of studying urban development, urban planning, and disaster management policies in relation to the vulnerability of affected populations in the selected study site, Istanbul.

The study is primarily concerned with the following questions:

---

[4] The International Society of City and Regional Planners (ISOCARP 1992) defined *planning*, in its *International Manual of Planning Practice*, as a "continuous process of thought, anticipating and preparing for foreseeable future." In order to manage such change in spatial terms, urban planning "makes arrangements for future demands on the use of public and private land," and seeks a balance between all interests "to resolve conflicting demands on space" (ISOCARP 2005: 50).

1. How does urban development and planning impact disaster vulnerability?
2. How much impact do natural disasters have on development and planning?
3. What are the factors that influence the adoption and implementation of disaster risk reduction in urban communities?

1.1.1 Research Design and Data Collection

The empirical research in this book was drawn on case studies in Istanbul, Turkey. Istanbul was chosen as a site, for it retained specific conditions that created the control variables for the research. Istanbul was affected by a major earthquake in 1999 and has a high probability of its re-occurrence. With a population exceeding ten million people, it is a megacity with dilemmas of a globalizing economy. It has extreme socio–economic and urban polarization allowing the study of disparities in urban and risk management.

This study was built-up of a two-stepped research, which was based on a combination of secondary data gathering, field, and survey research. The study began by examining the current hazard and risk profile of Istanbul, and then evaluated the role of urban planning actions and policies in leading to the city's vulnerability to disasters. As a second step, the study examined recent urban and risk management activities in the city following the 1999 Earthquakes.

Secondary data gathering included library and institutional research, as well as interviews with professional groups. Local data collection and consultations included research at the Department of Earthquake Engineering at Boĝaziçi University Kandilli Observatory, Center for Disaster Management at Boĝaziçi University, Urban Planning Program Library at Mimar Sinan University, the Economic and Social History Foundation of Turkey, Istanbul Branch of the Turkish Chamber of Architects, Istanbul Branch of the Turkish Chamber of City Planners, Atatürk Library of the Istanbul Metropolitan Municipality, Istanbul Branch of the State Institute of Statistics, and Istanbul Branch of the Ministry of Public Works and Settlement.

In addition, previously collected data for the *Columbia University's International Urban Planning Studio: Disaster Resistant Istanbul* study (CU 2002) was utilized; including research and consultations with the Istanbul Governorship Disaster Management Center, Geotechnical and Earthquake Investigation Department and the Master Planning Office of the Istanbul Metropolitan Municipality.

Field and survey research activities included development of a survey questionnaire to facilitate qualitative interviewing about disaster risk reduction activities carried out at local districts in Istanbul. A field research focused on observation of the urban context, as well as conducting interviews and collecting data in district municipalities in the city. Data collection and interviews at this stage included research at: Zeytinburnu and Tuzla District Offices of the

Governorship's Directorate of Civil Defense, Oyguç Civil Engineering Consultancy Firm in Kadıköy, Kadıköy Municipality, Zeytinburnu Municipality, Avcılar Municipality, Tuzla Municipality, and Pendik Municipality.

1.2 Theoretical Framework and Significance

As the planning theorist, Beauregard (2001: 438) argues, the diversity within planning is rooted in its "simultaneous occupation of multiple worlds." Planning is embedded not only with scientific analysis, but also with social relations and "the aesthetic concerns of the design professions" (Beauregard 2001: 438). This multiple occupation of the planning profession gives way to its interaction with various disciplines, leading it to have a very crucial role in managing disaster risks and reducing vulnerability, while at the same time creating socially and economically just urban environments.

The theoretical body of this book builds up on literature on natural disasters, vulnerability, and risk, and it links and extends the discussion to engage an understanding on urban systems.[5] This research and discussion adds to research and literature on urban, development, and vulnerability studies.

This book also promotes the significance of urban studies for sustainable development and disaster risk reduction. The study of sustainable development solely from a macroeconomic point of view may not help produce specific public policies to address the problems associated at the urban level. This book demonstrates the significance of socio–cultural and economic factors in disaster risk reduction, within the larger context of indicating that there cannot be one solution to mitigation, but that the larger framework should be tailored according to different socio–economic and physical contexts. The impact of socio–economic disparities in disaster risk reduction promotes the notion of sustainable development that concentrates on the well-being of all social groups without neglecting the urban poor who often accept higher hazard risks by occupying unsafe settlements. This notion of sustainable development is imperative in disaster risk and poverty reduction, particularly in developing countries, but is also applicable to other regions.

---

[5] The term *urban system* is used as identified by the United Nations (UN) guidelines on sustainable human settlements, as "both the largest unit capable of initially addressing the magnitude of urban, social, economic, political and environmental imbalances and the smallest scale at which problems can be meaningfully resolved in an integrated, holistic and sustainable manner" (UN 1996: 47–48).

1.3 Outline of the Book

The introductory chapter presents an overview of the book. It features the research questions and highlights the significance of study. Additionally, this chapter outlines research design and methods.

Chapter 2 provides a conceptual foundation for the book. It starts with examining global patterns in disasters and their impacts, and a discussion on the global trend of urbanization and climate change and their relation to natural disasters. This chapter continues with examining vulnerability in urban areas. It reviews the socio–economic, spatial, and institutional factors that result in disaster vulnerability in urban settlements, both in informal and formal urban areas. The chapter also examines various policies and projects that have been applied for disaster risk reduction. The chapter ends by investigating factors that influence the adoption and implementation of risk reduction measures, and their relation to urban vulnerability.

Chapter 3 provides a local case study to the examination of the interplays between urban development and vulnerability. It starts by examining Istanbul's current earthquake hazard and risk assessment. It continues with a historical analysis of Istanbul's urban development, planning activities and policies leading to its current socio–economic and physical vulnerability.

Chapter 4 examines urban and risk management activities that have taken place in Istanbul, following the experience of a major earthquake disaster in 1999. It also brings in municipality interview results examining the way risk and urban management are implemented in districts with different socio–economic conditions.

Chapter 5 brings together the discussions and research results presented in this book. As a summary and conclusion, it provides an overview and translates the interplays between urban development, risk management, and vulnerability from natural disasters, stressing the significance for good urban planning and good urban governance in disaster risk reduction and sustainable development of urban areas.

References

Beauregard R (2001) The multiplicities of planning. J Plann Educ Res 20:437–439

Centre for Research on the Epidemiology of Disasters (CRED) (2012) Disaster data: a balanced perspective. Cred Crunch 27

Columbia University (CU) (2002) International urban planning studio: disaster resistant Istanbul, final report (Unpublished). Urban Planning Program, Columbia University, New York

Gencer EA (2007) The interplay between natural disasters, vulnerability, and sustainable development. Ph.D dissertation, Columbia University, Urban Planning Program, Columbia

Intergovernmental Panel on Climate Change (IPCC) (2012) Summary for policymakers. In: Fields CB, Barros V, Stocker TF, Qin D, Dokken DJ, Ebi KL, Mastrandrea MD, Mach KJ, Plattner GK, Allen SK, Tignor M, Midgley PM (Eds) Managing the risks of extreme events and disasters to advance climate change adaptation: a special report of the working groups I and II and ii of the intergovernmental panel on climate change. Cambridge University Press, New York, pp 1–19

International Society of City and Regional Planning (ISOCARP) (1992) The international manual of planning practice. ISOCARP, The Hague

International Society of City and Regional Planning (ISOCARP) (2005) Four decades of knowledge creation and sharing. ISOCARP, Madrid

United Nations (UN) (1996) Guidelines on sustainable human settlements planning and management. Economic Commission for Europe, New York

United Nations (UN) (2011) Global assessment report on disaster risk reduction: revealing risk, redefining development. United Nations, Geneva

World Bank (WB) (2010) Natural hazards, unnatural disasters: the economics of effective prevention. The International Bank for Reconstruction and Development, Washington

World Commission on Environment and Development (WCED) (1987) Our common future. Oxford University Press, Oxford

Chapter 2
Natural Disasters, Urban Vulnerability, and Risk Management: A Theoretical Overview

> *Communities will always face natural hazards, but today's disasters are often generated by, or at least exacerbated by, human activities... At no time in human history have so many people lived in cities clustered around seismically active areas. Destitution and demographic pressure have led more people than ever before to live in flood plains or in areas prone to landslides. Poor land-use planning; environmental management; and a lack of regulatory mechanisms both increase the risk and exacerbate the effects of disasters.*
>
> Kofi Annan.

In his foreword to "Living with Risk," the United Nations' Secretary General, Kofi Annan[1] raised awareness to human-induced conditions that increase vulnerability to natural disasters. Rapid urbanization and land degradation, globalization and socio-economic poverty, global warming and climate change are among the global trends that affect the world at large and result in the severity, if not be the cause of natural disasters.

The increasing number and impact of natural disasters reveal themselves in statistics. The unprecedented rise in the number of natural disasters exposes a need to recognize global trends influencing this rise, and confront them through a larger policy framework. Furthermore, today as more than half of the world's population lives in urban areas, and coupling with the impacts of climate change, risk reduction in urban areas becomes more significant than ever. As the United Nations' *Local Governments and Disaster Risk Reduction* publication explains, "[u]rban risk, city planning and the role of local governments in dealing with risk reduction have been recognized as key factors to build resilient communities and nations" (UN 2010: viii).

This chapter starts by identifying the current state of global patterns of disasters and their impacts, and continues with examining the linkages between disasters and the global trend of urbanization and climate change. The chapter also studies vulnerability and risk reduction strategies in urban areas. The chapter concludes with discussions regarding the necessary elements for successful risk reduction in urban areas.

---

[1] Annan, Kofi, 2002: Foreword to *Living with Risk: A Global Review of Disaster Reduction Initiatives*, (UN/ISDR). Quoted in UN/ISDR, 2003: Disaster Reduction and Sustainable Development. A background paper for the World Summit on Sustainable Development; http://www.unisdr.org (2006):1.

E. A. Gencer, *The Interplay Between Urban Development, Vulnerability, and Risk Management*, Mediterranean Studies 7,
DOI: 10.1007/978-3-642-29470-9_2, © The Author(s) 2013

2.1 Disaster Patterns and Definitions

Worldwide statistics reveal the increasing number of disasters and disaster impacts within the last decades. Indeed, only within the last four decades, natural disasters have caused more than 3.3 million deaths and 2.3 trillion dollars in economic damages (WB 2010: 10). In the last three decades, two geophysical hazards, 2010 Haiti earthquake and the 2004 Indonesian earthquake and tsunami have caused the highest death toll from natural disasters. On the other hand, hydro-meterological hazards have been the dominant hazard types, affecting Asia, mostly, with tropical cyclones and floods, Africa with drought, and Europe with extreme temperature changes and heat waves (Fig. 2.1).

In the last three decades, it has been observed that many developing countries, especially those in Asia, have increasingly been impacted with aggregated disaster events causing an impetus in their development, such as with floods. Additionally, many developed nations have been impacted with single events in their hazard prone and increasingly exposed and vulnerable urban areas, such as experienced with 2005 Hurricane Katrina in the United States and the 2011 Japan earthquake causing immense monetary damages (US$ 210 billion and US$125 billion respectively). The variety in disaster typology, its distribution and impacts indicates the necessity to focus on different conditions of hazard, exposure and vulnerability and to produce strategic disaster risk reduction programs and policies (Fig. 2.2).

An increasing number of hazard and risk research and studies from different disciplines in earth, engineering, and social sciences have contributed to our contemporary understanding of disasters, vulnerability and risk management.

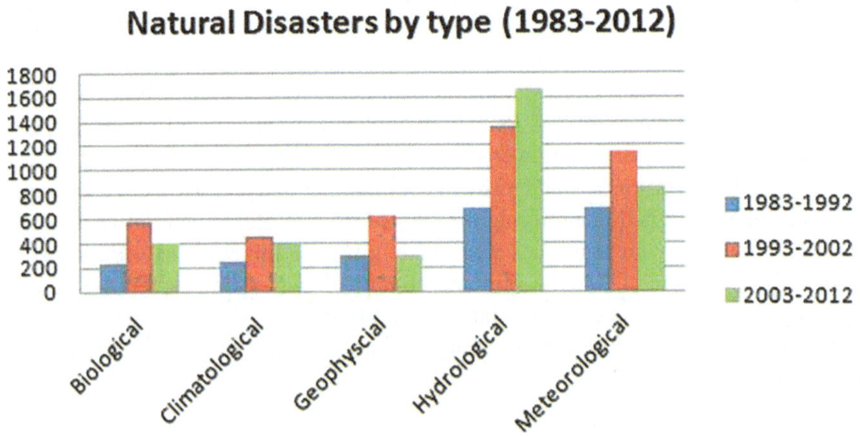

Fig. 2.1 Reported natural disasters by type (1983–2012) (by author). *Source* Raw data collected from EM-DAT: The OFDA/CRED International Disaster Database. Brussels, Belgium: Université Catholique de Louvain, Center for Research on the Epidemiology of Disasters (CRED) http://www.em-dat.net (Accessed 2012)

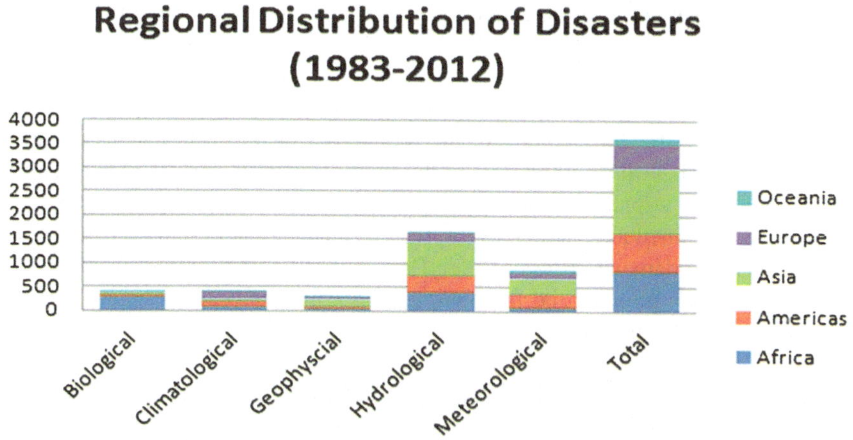

Fig. 2.2 Regional distribution of natural disasters, 1983–2012 (by author). *Source* Raw data collected from EM-DAT: The OFDA/CRED International Disaster Database. Brussels, Belgium: Université Catholique de Louvain, Center for Research on the Epidemiology of Disasters (CRED); http://www.em-dat.net> (2012)

However, this multiple exploration is based on different theoretical approaches and definitions of hazard, vulnerability, risk, and disasters. As Cutter (2001: 3) wrote, "the distinction between hazard, risk, and disaster is important because it illustrates the diversity of perspectives on how we recognize and assess environmental threats (risks), what we do about them (hazards), and how we respond to them after they occur (disasters)." While acknowledging these disciplinary differences, this book will use definitions of these terms provided by the United Nations International Strategy for Disaster Reduction Secreteriat (UNISDR).

Hazard is defined as "a dangerous phenomenon, substance, human activity or condition that may cause loss of life, injury or other health impacts, property damage, loss of livelihoods and services, social and economic disruption, or environmental damage" (UNISDR 2009: 17). In most cases, its origin defines the hazard, such as natural hazards or hazards that are induced by human processes.

Vulnerability is defined as the potential for loss (human, physical, economic, natural, or social) due to a hazardous event. It is the characteristics and circumstances of a community, system or asset that make it susceptible to the damaging effects of a hazard (UNISDR 2009: 30). Vulnerability encompasses the conditions determined by physical, social, economic, and environmental factors or processes, which increase the susceptibility of a community to the impact of hazards.

Exposure is "people, property, systems, or other elements present in hazard zones that are thereby subject to potential losses. Measures of exposure can include the number of people or types of assets in an area. These can be combined with the specific vulnerability of the exposed elements to any particular hazard to estimate the quantitative risks associated with that hazard in the area of interest." (UNISDR 2009: 15).

Risk is the possibility of harmful consequences or expected losses resulting from interactions between natural or human-induced hazards and vulnerable conditions. It is "the combination of the probability of an event and its negative consequences" (UNISDR 2009: 25). In the field of hazards and disaster research, risk is commonly expressed as the product of hazard, vulnerability and exposure.

Disaster is defined as a sudden event, such as an accident or natural catastrophe that causes great damage or loss of life. The UNISDR (2009: 09) defines *disaster* as "a serious disruption of the functioning of a community or a society causing widespread human, material, economic or environmental losses which exceed the ability of the affected community or society to cope using its own resources."

According to the UNISDR terminology, "disasters are often described as a result of the combination of: the exposure to a hazard; the conditions of vulnerability that are present; and insufficient capacity or measures to reduce or cope with the potential negative consequences. Disaster impacts may include loss of life, injury, disease and other negative effects on human physical, mental and social well-being, together with damage to property, destruction of assets, loss of services, social and economic disruption and environmental degradation" (UNISDR 2009: 09).

In the 2005 *World Conference on Disaster Reduction*, the current framework for disaster risk management was developed in the *Hyogo Framework for Action* 2005–1015: *Building the Resilience of Nations and Communities to Disasters* and was summoned as:

1. Ensure that disaster risk reduction is a national and a local priority with a strong institutional basis for implementation;
2. Identify, assess and monitor disaster risks and early warning;
3. Use knowledge, innovation and education to build a culture of safety and resilience at all levels;
4. Reduce the underlying risk factors; and
5. Strengthen disaster preparedness for effective response at all levels (UN 2005: 11–17).

This disaster risk management framework acknowledges the steps of traditional practice of disaster management (preparedness, response, recovery, mitigation), but also argues on giving attention to more "proactive strategies, which can contribute to saving lives and protecting property and resources before they are lost" (UNISDR 2004, 1:7). Emphasis is on risk reduction, which is defined as "the conceptual framework of elements considered with the possibilities to minimize vulnerabilities and disaster risks throughout a society, to avoid (by prevention), or to limit (by mitigation and preparedness) the adverse impacts of hazards, within the broad context of sustainable development" (UNISDR 2004, 2:3). This book will focus on "risk reduction" and "disaster risk management" through this framework and analyze urban risk reduction activities and disaster risk management in Istanbul in relation to sustainable development.

2.2 Urbanization and Natural Disasters

Today, more than half of the world population lives in urban areas making it essential to focus on urban areas for disaster risk reduction. The concentration of population and assets and the embedded conditions of socio-economic and spatial vulnerabilities generate disaster risk in urban areas affected by natural hazards. With the likely impacts of climate change, such as heat waves or elevation in sea-levels, today, exposure and vulnerability in urban areas deserve a special attention for disaster risk reduction.

Urbanization[2] and rapid population growth lead to the concentration of population in hazard- and risk- prone urban areas,[3] both in mega-cities[4] and in small- and medium- sized urban centers—although both types of urban growth represent different concerns for disaster risk.

While the majority of the urban population currently live in small- and medium-sized cities, this proportion is expected to grow at a slower pace. According to the 2011 *Global Report on Human Settlements* (UN-Habitat 2011), in 2000, 54.7 % of the world's urban population lived in cities of less than 500,000 people. This percentage is estimated to decrease to 50.4 by 2020. In contrast, while in 2000, only 8.2 % of the world's urban population lived in megacities larger than 10 million people, this percentage will increase to 10.4 by 2020; indicating the growing need to focus on rapidly increasing large and megacities for disaster risk.

The size, number, functions, and geographical distribution of medium- to large- and mega-cities create a major concern for disaster risk. In 1950, only 85 cities worldwide had populations of one million or more inhabitants. In developing countries, the number of these medium-sized cities increased six-fold since 1950. Today, there are 387 medium-sized cities, a big proportion of which are located in Africa, Asia, and Latin America (Fig. 2.3).

Due to the urban concentration of population, the greatest potential for disasters exists in the most populous cities. In 2000, the average size of the world's largest 100 cities was around 6.3 million inhabitants, increasing from 5.1 million in 1990, and from 2.1 million in 1950 (Wisner et al. 2004: 72, Satterthwaite 2005: 6). Over

---

[2] In simplest terms, *urbanization* is an increasing proportion of a population living in settlements defined as urban centers (Satterthwaite 2005: 2). The immediate cause of most urbanization is the net movement of people from rural to urban areas (which is mostly higher than urban to rural migration). It is important to note that national governments set their own population benchmarks to define what constitutes an urban area. Therefore, the scale of the world's urban population may vary according to different national standards.

[3] The proportion of people living in cities is lower than the proportion living in urban centers, as a significant proportion of people live in urban centers that are too small to be called cities (Satterthwaite 2005: 22). In this book, the term *urban area* will be used to identify both urban centers, cities, and their agglomerations.

[4] Mega-cities are cities with populations of ten million people or more. The United Nations first used the term in the 1970s to designate urban areas with populations of eight million or more. The threshold was increased in the 1990s.

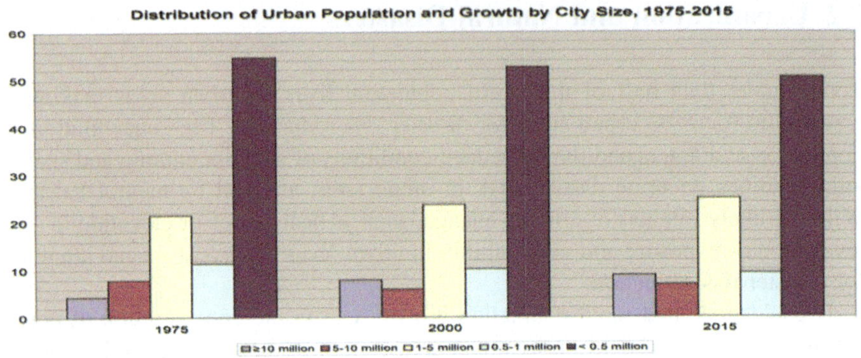

Fig. 2.3 Distribution of urban population by city size, 1975–2015 (by author) *Source* Data from United Nations Population Division (UNPD) (2002)

three-fourths of the one hundred largest cities are exposed to at least one natural hazard (UNISDR 2004: 1:59). Most of them are located in low- and middle-income nations and in hazard-prone areas particularly in Asia and in Latin America—a trend which is expected to continue within the next decade.

Mega-cities also bear major risks from natural disasters. According to data from UN-Habitat's (United Nations Human Settlements Programme) 2009 *Global Report on Human Settlements*, based on 2010 population estimates, there are twenty megacities[5] in the world. Ten of these megacities belong to low and lower-middle income countries, and the remaining ten belong to upper-middle and high income countries. Moreover, all megacities are exposed to natural hazards ranging from geological (earthquake ground shaking and mass movements) to meteorological (floods and storms) and climatic events (extreme heat and cold) and wildfires, indicating the necessity to think different risk reduction strategies for different conditions in megacities.

2.2.1 Urbanization and Climate Change

Climate change[6] is expected to increase hazard exposure and risks in many urban centers, particularly—but not only limited to—those located near coastal areas. Urban areas are expected to experience the effects of climate risk with rises in sea

[5] According to the 2010 UN population estimates, Paris, Jakarta, Kinshasa and Guangzhou (Guangdong) are other urban areas that will reach populations over ten million people by the year 2020.

[6] According to the Intergovernmental Panel on Climate Change (IPCC), *climate change* "refers to any change in climate over time, whether due to natural variability or as a result of human activity" (McCarthy et al. 2001: 3).

levels and the accompanying coastal floods and increases in the intensity and frequency of climatic events (Bigio 2003: 91), such as intense cold and hot events or intense rain and flash floods.

Among the most anticipated risks of climate change are the effects of sea level rise and accompanying hazards on small island states and coastal cities. According to the IPCC (2012), by 2100, while the global frequency of tropical cyclones will either decrease or remain unchanged, there will be, with the likelihood of 90–99 %, increases in the average tropical cyclone maximum wind speed and an increase in heavy rainfalls associated with tropical cyclones.

Sixty-five percent of the world's urban population currently live in coastal areas, and this percentage is expected to increase to seventy-four percent by 2025 (UN-Habitat 2011). Most mega-cities are either located on seacoasts or directly linked with riverbeds, increasing the exposure in hazard-prone areas. According to the IPCC (2012), by 2100, with the likelihood of 90–100 %, sea-level rise will contribute to upward trends in extreme coastal high water levels. Potential hazards in coastal areas and cities built near rivers are coastal flooding, erosion of beaches, sedimentation in river floors, flooding, and landslides. These hazards can intensify with a combination of intensified tropical storms.

In addition to these hazards, cities are also expected to be affected by severe heat and cold events. The Special Report of the IPCC (2012) projects that, during the twenty-first century, there will be, with the likelihood of 90–99 %, increases in length, frequency, and/or intensity of warm spells or heat waves over most land areas, and, with the likelihood of 99–100 %, increases in frequency and magnitude of warm days and nights at the global scale. Extreme cold events could lead to increase use of energy and worsening air pollution conditions, while expected heat waves could worsen in cities "pronounced as heat islands" due to the heating up of the concrete buildings and paved areas.[7]

In their summary report, the IPCC[8] (2001) stated that, "the developing countries, particularly the least developed countries have lesser capacity to adapt and are more vulnerable to climate change damages, just as they are more vulnerable to other stresses," and continued that, "[t]his condition is most extreme among the poorest people." Climate change is expected not only to alter the intensity and the frequency of hazards, but also to increase the vulnerability of societies, requiring a special attention to the study of disaster risk reduction in urban areas.

---

[7] Munich Re Group 2005: Megacities—Megarisks: Trends and Challenges for Insurance and Risk Management. Munich Re Group Knowledge Series; at: http:www.munichre.com (2006):25.

[8] Intergovernmental Panel on Climate Change (IPCC) (2001).

2.3 Vulnerability and Risk Reduction in Urban Areas

Disasters vulnerability in urban areas arises from a result of a combination of interrelated physical, socio-cultural, economic, and institutional conditions. The buildup of exposure due to concentration of population and assets, increased susceptibility due to physical condition of buildings or infrastructure, social and economic composition of residents, and lack of institutional capacity result in disasters in hazard-prone urban areas.

In urban areas, there is a strong tie between vulnerability and urban poverty,[9] and an understanding of urban poverty encompassing both economic and non-economic factors provides insight to disaster vulnerability. On the other hand, it is necessary to stress that *vulnerability* is not identical with *poverty*; and that "not all poor people are vulnerable to disasters, and some people who are not poor are also vulnerable" (Bankoff 2003: 19). This section examines vulnerability in urban areas in two sections: (a) in informal settlements in mostly peri-urban areas and (b) in formal settlements in core cities, with the understanding that there are many overlapping elements of susceptibility in both areas and the differences between the two are increasingly disappearing, especially in the fast growing megacities in developing countries. The aim of these discussions is not to focus entirely on what is vulnerable, but also to discuss who is vulnerable and why, and to explore risk reduction strategies.

2.3.1 Vulnerability in Informal Settlements

Within the last decades, population shifts from impoverished rural economies, pressures of globalization and industrial relocation in major cities have contributed to one of the biggest urban challenges in developing countries: the expansion of urban areas and the creation of unplanned informal settlements as the sole option for newcomers. Even though informal settlements,[10] squatters, and slums have

[9] The World Bank (WB) defines poverty as an unacceptable deprivation in human well-being; which goes beyond the traditional view as measured by income or consumption; but that includes basic material needs including adequate nutrition, health, education, and shelter as well as social needs including security and empowerment (WB 2001; Ames et al. 2002). According to the World Bank's Poverty Reduction Strategies, urban poverty is explained with dimensions of income poverty, health and education poverty, personal and tenure security, and disempowerment (Baharoglu and Kessides 2002).

[10] Informal settlements have recently been defined and used under the large umbrella of the term *slum*. Standard and operational understandings of slums include both its traditional definition as declining housing areas that have deteriorated with the movement of their original dwellers to new and better areas of the cities, as well as informal settlements in urban periphery of mostly developing nations and that encompass both squatter settlements and illegal subdivisions (UN-Habitat 2003: 9). In this book, the terms *slum* and *informal settlements* are used interchangeably.

long been in existence, these settlements have grown in numbers and in spatial forms with the increase of the urban poor and their exclusion from formal housing sectors. In many cases, with urban spatial growth, formerly independent administrative and political units of settlements have been incorporated to metropolitan cities, creating peripheral municipalities and generating new challenges in urban governance.

Another impact of these migratory practices has been the reduction of the rural–urban relationships for livelihood, as the expansion of urban areas to fertile urban land has resulted in the reduction in the food supply of urban residents, increasing urban poverty and vulnerability. In *Sustainable Land Management*, Hari Eswaran and his co-authors (2011) write about the effects of the mass migration from Eastern and South-Eastern Turkey to Seyhan basin in search of jobs in irrigated plain and explain that these migratory practices and enlarging urban occupation has effected fertile soils of the delta exceeding "the settlement urban/rural farmland and the natural environment" ratios of the legislation developed for the sustainable management of this land. The disruption of agricultural production and related livelihoods by the expansion of urban land markets not only increases poverty and food insecurity, but also creates serious future climate problems with the loss of land surface necessary for the water-cycle[11] or environmental problems with soil erosion contributing to the silting up of drainage channels and consequently increasing vulnerability of residents who migrate from rural areas and settle in these land (Sattherwaite/Tacoli 2002: 52–70).

Along with conditions of urban poverty, informal economy, and challenged urban management systems, informal settlements and their residents have become increasingly susceptible to natural disasters. Statistics indicate that just in Latin America and the Caribbean, which is highly prone to a variety of natural hazards, 27 % of the urban population live in slums,[12] with some countries this percentage is much higher, such as in Nicaragua with 45.5 %, and Haiti with 70.1 %, indicating the increased risk in these settlements.

First, most informal settlements carry physical vulnerabilities due to their location or construction practices. These settlements are often "located on land not deemed appropriate for habitation because of its steep terrain or geological characteristics that make it prone to subsidence, landslides, or mudslides" (UN-Habitat 2003: 69). Slum dwellers and squatters often settle in these dangerous locations as the only option for their livelihoods and survival. An example is the large squatter settlement in Central Delhi that has "existed within the designated flood plain of the Yemuna River for more than 25 years" (Sanderson 2000: 98). According to David Sanderson (Sanderson 2000: 98), "[t]he settlement is forced to evacuate at least once a year to the busy roadside whilst their shelters are flooded for upwards of one month. The regular flooding is seen as the price to pay for living in the centre of the city at low cost." In Belize, where the slum population is

[11] Communication with Prof. Dr. Selim Karpuz.

[12] Dodman et al. (2009).

equal to nearly half of the urban population, the "low-lying coastline accommodates approximately 45 % of its total population in densely populated urban areas such as Belize City," and "[t]hese coastal centers represent some of the country's most vulnerable to storm events as they lie approximately one to two feet below sea level" (WB/GFDRR 2010: 94).

On the other hand, many times, environmental degradation, loss of rural incomes and strict building codes lead the incoming populations to the only available land, to the risk-prone urban fringes. For instance, situated between the Pacific Ocean and the Andes, Lima is subject to floods, mud and landslides, and it is prone to earthquakes. With the Pan-American Highway linking Lima to other port cities, rapid urbanization along the coastline has contributed to increased levels of risk (UNISDR 2004: 1:60). Within the last decades, in addition to the city's coastal growth, informal squatter settlements have proliferated around the fringes of Lima in unstable alluvial soil along the riverbanks or in hillsides (Oliver-Smith 1999: 248–294). Perlman (1993: 34) has argued that "counterproductive incentives" have increased the informal housing sector in this Latin American city. Perlman (1993: 34) explained that in Lima, "[t]he average period needed to acquire a house formally is nearly 7 years; to obtain a land title takes 31 months, and to secure a construction permit takes another 12 months. Thus, the vast majority of low-income families are forced into the vulnerable position of having to find housing 'informally,' without minimal legal protection." Oliver-Smith (1999: 273) has written about the development of these settlements in Lima: "During the 1950s, there were 56 such settlements located on the periphery of the city; in 1984 there were 598 such *barriadas*. Now called *pueblos jovenes*[13] (young towns), they contained close to 40 % of Lima's population. Older *barriadas* gradually evolved into permanent communities and grouped together to form separate municipalities." Similar patterns of vulnerability are reported in Manila, where "informal settlements at risk of coastal flooding make up 35 per cent of the population; in Bogotá, 60 % of the population lives on steep slopes subject to landslides; and in Calcutta," where "66 % of the population live in squatter settlements at risk from flooding and cyclones" (Pelling 2003: 28).

Inadequate building materials accompany risk by physical exposure in squatter settlements as structures are often built with non-permanent materials, such as "earthen floors, mud-and-wattle walls or straw roofs" (UN-Habitat 2003: 11). Quick makeshift structures are observed in impromptu urbanizations and sprawls of many low-income countries. For instance, the case of Mumbai's (Bombay) sprawl is attributed to the city's shift of its industrial base from import substituting to export orientation, and relocation of industry from central city to highways extending to periphery (Pelling 2003: 29). In his exploration of postmodern Bombay, Jim Masselos (1995: 212) wrote: "A global city like Bombay is in fact

[13] The popularisation of *pueblos jovenes* in official terminology, instead of the former term of *tugurios* (inner-city slums) *and barriadas* (squatter communities), is argued to be an attempt of authorities "to address the damaging effect of prejudice against slums" (UN-Habitat 2003: 10).

predominantly a village, a series of villages represented in the shanty structures that permeate the city. Shanty structures derive from village prototypes in rural India but are modified by the requirements of space and the availability of materials—plastic, tin, bits of cloth, wood and bricks, which draw on past and present materials." Indeed, according to the 1991 census in Mumbai, 60 % of registered buildings in the city are "informal masonry and other non-engineered buildings of light material used in slum areas"[14] (Wenzel and Bendimerad 2002: 117). According to the Government of Maharashtra, vulnerability of these buildings is "so bad that shaking with intensity VII is expected to significantly damage 50–75 % of them" (Wenzel and Bendimerad 2002: 117).

Most makeshift squatter settlements built with impermanent or recycled materials belong to the newcomers or to the very poor. In many cases, these settlements lack municipal services and infrastructure. For instance, a household survey carried between inter- and intra-urban entities in São Paula, Accra, and Jakarta in 1991 found out the following results in the poorest 20 % of the populations: 67 % of the poor in Accra, 31 % in Jakarta, and 19 % in São Paula had no water source at residence; 69 % of the poor in Accra, 32 % in Jakarta, and 7 % in São Paula had to share toilets with more than 10 households; and 97 % of the poor in Accra, 52 % in Jakarta, and 14 percent in São Paula had no home waste collection (McGranahan et al. 2001: 67–83). Likewise, in Nicaragua, with 45.5 of slum population, only 52 % has access to improved sanitation, and in Anguilla with 40.6 % slum population, only 60 % of the population has improved drinking water sources showing the high degree of vulnerability due to lack of infrastructure in informal settlements.

Lack of proper infrastructure facilities and unplanned urbanization schemes combine to create new hazards in informal settlements, where inadequate waste disposal in riverbeds and ravines, in addition to the urbanization of watersheds and wetlands may modify hydraulic regimes. This is the case in Quito, Ecuador, where with pressure of unplanned urbanization, approximately 3.2 kilotons of solid waste is disposed of in ravines each year, obstructing drainage and increasing flash flood hazard.[15] Similarly, Kante (2005) reports that in the capital of Uganda, Kampala, the expansion of the city into the wetlands through slum building, and the dumping of waste into these wetlands and surrounding canals has resulted in several floods, as these wetlands had previously served to store water for the city.

As informal settlements grow larger and denser, lack of sanitation, clean water and garbage removal, add congested living conditions add to the disaster vulnerability of slum dwellers, resulting in further environmental and health problems. The UN Millennium Task Force on Slum Dwellers reports that lack of provision

[14] It should be noted that shanties or slums in Mumbai are a combination of peripheral and inner-city settlements. Indeed, one of these inner city squatter settlements, Dharavi, which was the largest slum in Asia in the 1980 s, has a population estimated to be somewhere between 500,000 and 1 million people, but today there are four other slums in Mumbai larger than Dharavi. "Dharavi in Mumbai is no longer Asia's largest slum", in *The Times of India* (6 Jul 2011).

[15] United Nations Development Programme (UNDP) (2004): 61.

Fig. 2.4 Sidewalk shacks built with cardboard, tin and wood material in Jamaica (Photograph by author, 2012)

for water and sanitation and high levels of overcrowding contribute to many communicable and non-communicable diseases (from respiratory infections to malaria), injury, and premature deaths (from rapid spread of vaccine preventable diseases) in several urban slums in Dhaka, Nairobi, and São Paula (UN Millennium Project 2005: 59–60). In the Dominican Republic, where 17.6 % of the population is slum dwellers, and the proportion of the population using improved drinking water sources and improved sanitation facilities are 86–83 % respectively, "[t]he health status of the population influences vulnerability," with food or water-borne, water contact or vector borne infectious diseases (WB/GFDRR 2010: 129). Indeed, in the Cental American Countries, where there is a high rate of urban slum dwellers, estimated mortality rates for infants less than age 1 is very high; for instance in the Dominican Republic, 46; in Nicaragua, 40; and in Haiti 87 deaths occur per 1,000 births.[16]

In many informal settlements and peripheral municipalities, vulnerability to natural disasters does not end with such physical exposure or social fragility. Lack or inefficiency of public urban services and institutions—transportation networks, hospitals, fire- or police stations—translate into *lack of response capacities* at times of disasters. Informal land titles obtained through developers add to the limited disaster recovery of these settlers, who can neither obtain government aid nor credit with their illegal titles. Social exclusion, ethnic or immigrant status, poor education and limited job opportunities add to the income poverty of these residents, limiting their mobility and resettlement and creating one of the biggest challenges for urban policy making in the developing world (Fig. 2.4).

[16] Dodman et al. (2009): 29–30.

2.3.1.1 Risk Reduction Strategies in Informal Settlements

Risk reduction strategies for informal settlements ensued the way these settlements have been perceived by officials, whether they were international development agencies or local public administrations. Many scholars describe that general attitude towards informal settlements, slum dwellers, and squatters in developing countries have usually varied from "blind intolerance to blatant hostility" (Westgate 1981: 28) by local officials in charge with urban management, who considered these settlements as a "cancerous growth on the city" (Laquian 2005: 353).

Starting in the 1950s, programs attending to the problems of these settlements focused on their eradication by bulldozing and evictions. In their exploration of the housing problems in the Third World, Jorge Hardoy and David Satterthwaite (Hardoy and Satterthwaite 1993: 111–160) summarize government justifications for these evictions in three categories: (1) city beautification programs; (2) slums as centers of crime and health problems; and (3) redevelopment for public projects.

In rare cases, these demolitions have also been targeted towards specific groups, whether be by ethnic marginalization or by political agenda. In Zimbabwe, in a slum demolition campaign in 2005, seven hundred thousand people were left homeless in what was called by the government an "urban clean up effort," but what, according to human-rights activists, was aimed at peasants, who made up the core of the political opposition to President Mugabe's rule (Wines 2005).

In some cases, as in the situation in Seoul between 1983 and 1988, despite the fact that the government had destroyed about 48, 000 buildings to host the Olympic Games, only a very small portion of the evicted people received new accommodations[17] (Hardoy and Satterthwaite 1993: 118). In other cases, these demolitions were accompanied with redevelopment or re-housing projects, which attempted to resettle population at "considerable distances" from the city to *superbloques* of public housing, such was in the case of Venezuelan evictions of the 1950s (Hardoy and Satterthwaite 1993: 118). However, most cases of resettlement approach have been unsuccessful, for they have only transferred the problems of the urban poor to other locations without providing amenities and employment opportunities, and at times destroying the important kinship ties that many of the migrants share and connect to. Many governments have stopped using the resettlement approach as a first strategy after criticisms and the involvement of international development organizations. For instance, Laquian (2005: 354) reported that "[i]n the Philippines, it was mandated by law that people can be moved from a site only if (1) they are staying in dangerous places such as riverbanks, steep slopes, along railroad tracks, or near toxic waste dumps; (2) the occupied land is

---

[17] A similar slum clearance campaign is experienced today in Rio de Janeiro as the government is preparing for the 2016 Olympic Games and pushing out drug gangs in the favelas and these slums are now turning into lucrative real estate opportunities for the wealthy.

needed for an infrastructure project that is required for the general welfare; or (3) an occupant is in clear violation of another person's property rights."

Increasing awareness of right to housing, as was strongly established with the 1996 Habitat Agenda, and the failure and criticisms of the repressive eviction strategies of local and national governments led to new strategies to deal with the living conditions of slum dwellers. Beginning in its earliest period in the 1970s, self-help and in situ slum upgrading policies were based on the concept that "urban poor have the capabilities to effectively deal with their own housing problems," and that, "given such assurances as security of land tenure, low interest loans, appropriate building materials, and some technical assistance," they could help upgrade their own living conditions (Laquian 2005: 362). These projects and policies focus on three main areas of concern: (1) provision of basic urban services; (2) provision of secure tenure for slum dwellers and the implementation of innovative practices regarding access to land; and (3) innovative access to credit (UN-Habitat 2003: 130).

Slum upgrading projects have proved to show success in their early stages. For instance, Indonesia's Kampung Improvement Program "upgraded existing low-income communities by improving roads and footpaths, drainage, flood control, water supply, communal toilets, and garbage collection and disposal," and the project was expanded into a nationwide effort (Laquian 2005: 363). The results of the program showed that households in project invested twice as much in home improvements than other households (UN-Habitat 2003: 130).

Another well-known project, the Orangi Pilot Project, was organized in the largest *katchi abadi* (informal settlement) in Karachi. Between 1980 and 1992, the project improved water, sanitation, and sewerage facilities through voluntary community action, benefiting about one million people. Due to the success of this project, four other community organizations carried out similar projects in Karachi; and the emphasis expanded to include building material provision, small-scale credit and livelihood improvement (Laquian: 205, 363). However, the sustainability and the success of this project could not be accomplished when applied in other communities in Pakistan. Laquian explains that failures of these applications had come from inadequate provision by municipal networks for the connections of the self-built coverage. The United Nations Human Settlements Program (UN-Habitat) reports other slum upgrading projects, in which inadequacy of municipal provisions had brought failure, when "[g]overnments did not follow through with services, communities did not maintain the facilities, and governance structures disappeared once the international experts were gone" (UN-Habitat 2003: 131).

Today, there are several initiatives to evolve slum upgrading and resettlement programs into more sustainable and integrated development approaches. In Tunisia, over a 30 year period, both the national and city governments shifted their approach from "slum" clearance to "slum" upgrading, and their focus from bringing infrastructure and public amenities to a long-term policy of supporting the development for land for housing for non-poor groups, which increased overall supplies and reduced costs eliminating the need for informal settlements.[18]

What is called as "participatory slum improvement," "integrated slum upgrading," or "urban upgrading" projects aim at having a more holistic approach to slum upgrading and risk reduction by considering problems of communities as a whole, involving both governments and communities, and requiring empowerment of communities, in addition to financial stability and commitment of local administrations. The *Global Report on Human Settlements* states the more sustainable efforts in slum upgrading efforts to be those "that are the main plank of city development strategy with planned, rolling upgrades across the city and a political commitment to maintenance" (UN-Habitat 2003: 132). It argues that "[a]s a general rule, the more marginalized or culturally separate the group being assisted, the more participation and partnerships are necessary" (UN-Habitat 2003: 132).

The Mumbai Railway Dwellers Resettlement Project is one of the projects that have required the empowerment of the community and the involvement of the local government. This project led to the participatory resettlement of ten thousand families, who lived adjacent to the railway tracks in Mumbai, into accommodations with assurance of secure tenure and basic amenities of water, sanitation, and electricity within one year of their negotiations with the Maharashtra Government. This negotiation was made possible with the empowerment of the community through the self-organized *Railway Slum Dwellers Federation* that was aided by a non-governmental organization, SPARC, and with the transference of power from government agencies in charge with resettlement and rehabilitation to the NGO alliance (WB 2003: 125).

In *The End of Poverty*, Sachs (2005: 240–241) detailed the continuing positive effects of this alliance and the project on slum dwellers in Mumbai: "...group action has taught them that in fact they have legal rights within the city and even the possibility of access to public services if they act together... With SPARC's initiative, the new Slum Rehabilitation Act has given added power to the communities: slum-dweller organizations are now legally empowered to act as land developers if they can demonstrate that they have agreements to represent at least 70 % of the eligible slum dwellers in a particular location," and they "can tap into special municipal programs to gain access to real estate for community resettlement or for commercial development that can finance resettlement elsewhere."

Another project that facilitated slum dwellers' involvement used an integrated approach to vulnerability reduction in their high-flood risk communities in Mozambique. Developed by the UN-Habitat, this integrated slum upgrading and vulnerability reduction project aims at strengthening relationships between central government, local authorities, and resident communities. The project was promoted under the *Cities Without Slums Initiative* and included three main components: (1) support policy-making; (2) training and capacity building; and (3)

[18] Dodman et al. (2009).

participatory land use planning and physical implementation at the local level (Spaliviero 2006: 106–115).

As a first step of policy-making support in Mozambique, where there has been an absence of regulatory instruments and coordinated institutional frameworks related to urban planning, the UN-Habitat co-founded the preparation of Territorial Policy Law Project in order to "set the legal framework of reference regulating all physical activities and coordinating existing laws" (Spaliviero 2006: 108–09). Mathias Spaliviero, from UN-Habitat at Mozambique wrote about the project and explained that, in addition to "strengthening the urban management technical capacity at the local level by placing skilled national professionals in the Municipalities," this project advocated for "the active participation of the community in the planning process," and argued that "preparedness and mitigation techniques could minimize the negative impacts of moderate flooding" (Spaliviero 2006: 109). With that, a training program, *Learning how to live with floods*, was launched in 2003. Through this awareness program, the project aimed at educating different parts of the society to issues such as "factors causing the floods, type of flood risks, different preparedness and mitigation techniques, contingency planning, community self organization, response actions" (Spaliviero 2006: 110).

As a final step of participatory land use planning, local area consultations were made confirming problems in informal settlements such as poor drainage efficiency, difficult access to safe drinking water, lack of sanitation facilities, inadequate road network, and inefficient waste management. In informal settlement areas in four cities, Maputo, Chókwè, Tete, and Quelimane, land use and disaster management plans with priority intervention and methodological instructions for slum upgrading strategies were introduced. Spaliviero argues that the involvement of the central Government, local authorities, and local communities has provided a trust and strengthening of the relationships. In Quelimane City, "under the supervision of the municipal technical staff and the coordination of a local committee, almost 400 dwellers were contracted on a rotational basis during a period of two months" (Spaliviero 2006: 113). The community selected to clean and regularize a cumulative drainage channel, and to improve their main access road that flooded after each rain event. Spaliviero describes that "[t]his positive experience has reinforced the community's will to contribute to improving the living conditions of their own neighborhood" (Spaliviero 2006: 113). However, as previous slum upgrading projects have shown, for the sustainability of this and other projects, and their long-term maintenance and upgrading, there should be stable support from the involved agencies, both internationally and nationally, as well as a strong institutional framework, which would capacitate local governments and provide legal rights to slum dwellers.

These provisions are addressed in the evaluation of a much broader integrated urban upgrading project of the World Bank, in Riberia Azul, Salvador, Brasil.[19] This program in a low-income neighborhood in Salvador, Bahia covered forty

[19] Baker (2006).

thousand families, representing 6 percent of Salvador's municipal population. This area was characterized as "'high-risk' situated in a flood prone area, with a large number of squatter settlements, insecure land tenure, a highly polluted environment by household and industrial waste, poor social indicators, and very limited access to infrastructure and basic services".[20] The program combined physical interventions with investments to improve the social and economic conditions of the population. Projects included "housing and infrastructure improvements, and programs in health care, child nutrition, education, training, and employment generation through cooperatives".[21] Community participation has been a fundamental part of this project, which was implemented by CONDER (Urban Development Company of the State of Bahia), AVSI (Association of Volunteers for International Service), and an Italian and a local non-governmental organization (NGO) partnership.

This urban upgrading project introduced housing and infrastructure works including improved access roads, storm drainage, water supply and sanitation, solid waste collection, housing improvements, and resettlement of those living in risk areas, particularly in the *palafitas* (stilt houses informally constructed over the inlet). An evaluation of the project showed that residents reported several positive benefits of housing and infrastructure improvements. However, due to hard terrain conditions, the heavily engineered new housing costs were found considerably more expensive than building new units in available plots, bringing forth the "scope for a policy shift towards providing inexpensive serviced land and access to credit rather than housing".[22]

Other lessons learned from this pilot upgrading project were reported as: (a) capacity building for community associations can be highly beneficial; (b) environmental planning for individual community needs to be integrated with a broader systemic plan at the city and state level; (c) strengthening inter-governmental relations could improve service delivery; (d) clear roles and responsibilities, as well as their flexibility are needed in institutional arrangements; (e) participation is critical to successful implementation and sustainability; and (f) municipalities will need to play a greater role from the start, particularly to ensure program sustainability.[23]

As observed in previous examples, the results of the Salvador slum-upgrading project indicates the significance of partnership and secure relationship between all levels of involvement from central government to local authorities and empowered local communities for the sustainability of integrated slum upgrading and risk reduction programs in informal settlements.

---

[20] Baker (2006): 1.

[21] Baker (2006): 2.

[22] Baker (2006): 23.

[23] Baker (2006): 23.

2.3.1.2 Vulnerability in Formal Urban Areas

Physical exposure to disasters is not a condition that belongs solely to the very poor, nor does it need to transfer into risk. In many cases, adequate building standards and urban planning actions alone can help manage or reduce disaster risk by physical exposure. However, these actions have been absent, or when available, not properly applied in many "formal" urban areas.

This section will explore this phenomenon in concrete urban agglomerations of the post-1950 era, encompassing the first generation building boom with concrete framed apartment buildings and the post-1980s building boom including *vista communities*, referring to vacation homes in coastal developments and hillside or seaside residences.

Initial stages of modern concrete agglomerations were in societies that previously used traditional building materials and architectural styles. Starting in the 1950s, the process of modernization coupled with rural to urban migration and initiation of private building activity changed urban landscapes in many countries. Spontaneous settlements proliferated around major cities of the Mediterranean Europe, be it Barcelona, Rome, or Naples, while many cities experienced the destruction of existing housing stock and the construction of apartment buildings.

Writing about the period from 1951 to 1981, Leontidou (1990: 142) argued that building process in this era had "erased the neo-classical architectural tradition" of Athens. Leontidou wrote, "Greater Athens was subject to an aggressive invasion of capitalism, and was changed into a reinforced concrete agglomeration, where building space was commercially exploited to the maximum degree possible. The multi-storey apartment blocks were constructed in a piecemeal process within a fragmented housing market. Most of them were low-quality constructions. Building standards declined, with the result that a large proportion of recently built housing in Greek cities is already in need of repair or even replacement" (Leontidou 1990: 142–44). During the same decades, similar style of building activity was also prevalent across the Aegean, in major cities of Turkey, where old housing stocks were being destroyed while apartment blocks were being built. Today, the seismic Southern Mediterranean cities still consist of the housing stock of the early modern concrete era. However, despite lack of adequate building regulations at the time, in many instances quality of housing constructions of this period have proved to be higher than that of post-1980 building boom.

In many developing or middle-income countries, two distinct types of housing stock may represent most of the post-1980 agglomerations. The first type is primary housing of the low- to middle-income groups in major cities of growing economies. These are housing responses to rapid population growth with higher quality material use than slums in the low-income countries, but with similar problems of physical vulnerability. The second type is primary or vacation homes of the middle and upper-middle income groups, a housing model observed from coastal cities to hillside residences.

Problems in both development types usually start with an increase in building activity with an unqualified construction sector. Oversight of control due to inadequacy or corruptions of local governments and officials add to the problem. For instance, in the touristic Caribbean Islands of Grenada and St. Vincent and the Grenadines, "[n]ew construction, particularly in relation to tourism, continues with little formal land use planning or construction code enforcement," as the construction codes that exist are not evenly applied (WB/GFDRR 2010: 161, 239). Likewise, "[p]oor regulated construction and land use practices" are found to be "among the biggest contributors to risk from losses" in the Island of Saint Lucia, where "[l]ack of uniform enforcement of building codes contributes to the vulnerability of island infrastructure (WB/GFDRR 2010: 229). In other cases, non-adequate applications of building codes or deficient structural configurations are the main cause of vulnerability. In Panama, which has one of the larger urban settlements in the Central America and the Caribbean region, "[t]he poor enforcement of national and local land use regulations, the uncertainty about compliance with building codes, rapid demographic growth and unplanned urban and industrial expansion" are found to be "responsible for most of the current and significant increases in vulnerability" signifying the susceptibility of populations and assets at the wake of loose enforcement or building code and regulations (WB/GFDRR 2010: 21). Many times, structural configurations are executed after the completion of buildings, as residents try to reconfigure their living spaces without consultation to architects or civil engineers (Figs. 2.5, 2.6).

Recent earthquakes have revealed that modern constructions in many urban areas lack basic earthquake resistant characteristics, even though design codes and

Fig. 2.5 Shoring of balconies in vacation homes in Playa d'Aro, Spain (Photograph by author, 2003)

Fig. 2.6 Scaffolding in Kathmandu, Nepal (Photograph by author, 2013)

building standards have been updated to provide safety of structures. The *Reconnaissance Report* of the Earthquake Engineering Research Institute (EERI 2003) for the 2003 Boumerdes, Algeria earthquake concluded in similar observations. The reconnaissance team examined the destruction caused by the earthquake epicentered in the province of Boumerdes, east of the capital city of Algiers. According to the report, the heavily damaged two areas had undergone different urbanization processes. The first of the damaged areas was in Algiers, where destruction had occurred mostly in new structures, and a result of the changes in the State's role in construction sector and planning system. In the 1990s, as Algeria was transforming from a rigid-state controlled system to a free-market economy, the State made major changes in planning and construction regulations. With the liberalization of construction regulations, an unqualified private sector emerged, hastily developing housing mostly with government oversight and without building permits. According to official data, in Algiers and its vicinity, "in the period during 1990–2002, 42.4–52.8 % of the individual homes were built without a legal title document, and thus without a building permit" (EERI 2003: 5). Most of these developments were along the coastal districts with high real-estate value. The reconnaissance team argues that corruption and personal interventions had interfered with the attention to the quality of construction, resulting in heavy damage to this housing stock (EERI 2003: 3–11).

On the other hand, urban development in Boumerdes had taken another path. The city was created in 1958, as part of a "French economic reform plan for

Algeria," and it was "intended to serve as an administrative and educational outpost" (EERI 2003: 6). The EERI (2003: 6–7) team reports the emergence of three generations of buildings during the planning of Boumerdes: "the first generation of buildings, built between 1959 and 1974, has bearing walls; the second generation, built between 1974 and 1993, is primarily engineered multistory buildings built by large government-owned construction companies; and the third generation is characterized by a return to traditional architectural methods using reinforced concrete beam-column and concrete slabs with brick partitions." According to the EERI report, damage in Boumerdes had mostly occurred in institutional[24] and large scaled apartment buildings (EERI 2003: 9).

The 1985 earthquake in Mexico showed similar destruction patterns to what was experienced in Algeria. According to Meli (1993), in Mexico "buildings constructed before 1950, with flexible, inadequately detailed, and almost unconfined concrete elements, have performed, in several instances, better than those with modern construction." Meli and Alcocer (2004: 31) attribute this situation to the replacement of the thick infill and façade masonry walls with lighter and weaker partition elements without updating the detailing rules of the 1950s. Therefore, they explain, "the poorly detailed modern reinforced concrete frames exhibited more severe earthquake damage than older frames with equally poor detailing but with more substantial nonstructural elements" (Meli and Alcocer 2004: 33).

In the same earthquake, a second set of damages was recorded in mostly government-sponsored projects. Documenting the impact of that earthquake, Puente (1999) wrote that 30 % of the government hospital capacity in Mexico City was lost with the earthquake, and that most of these buildings were post-1950s constructions. According to Puente, one of the biggest damaged residential areas was the Nonalco-Tlatelolco housing estate, which was comprised of 102 separate buildings. The estate was constructed in the early 1960s, and it "was intended to be a model of state responses to joint needs for slum clearance, new housing, and improved architectural design" (Puente 1999: 306).

In assessing these damages, it is also essential to consider geologic conditions of the location. In his famous textbook on earthquakes, Bolt (2004: 279) wrote that due to considerable distance between the earthquake source and the Valley of Mexico, "few structures built on firm soil and rock suffered damage." On the other hand, one area near the city center that was "underlain by a thick deposit of very soft, high-water content sands and clay" encompassed "most of the buildings that collapsed" in the 1985 earthquake (Bolt 2004: 280).

---

[24] Meli and Alcocer (2004: 33) explains that "the rate of distress and failure suffered by school and hospital buildings after major earthquakes is consistently higher than, or at least equal to, than of other common buildings." They argue for the existence of two major reasons for this high rate of damage. One is related with "inconsistency among design seismic-induced loads, expected performance, and design and detailing rules," and "the second reason is related to the more complex and irregular structural layouts" of these buildings (Meli and Alcocer 2004: 33).

Decision to build in geologically unstable or high-risk areas is a matter of available land for developers, to locate in these buildings is a matter of economics for most urban residents, but for those who wish to live in the most scenic areas, it is a matter of choice that is made with or without adequate information. Indeed, today in many developing countries, there are examples of high-income groups living with the same informality as slums or squatter settlements,[25] in scenic areas that are not open to settlements due to their protection or high-risk of hazards. For instance, in his research on ecological sustainability in Mexico City, Pezzoli (1998) explains that in Mexico City's green-belt zone of Ajusco, an area declared for ecological conservation in the planning departments of the city, rural land has transformed into an urban land since the 1970s. Pezzoli writes that this was not only a result of low-income groups' settlement, but also of real-estate developers and higher-income groups attracted by the "zone's greenery, clean air, and panoramic vistas" (Pezzoli 1998: 194). Pezzoli records the contradictory enforcement of zoning laws in this area by public officials favoring development of higher-income groups, while at the same time taking steps to eradicate irregular settlements with arguments about their negative impacts on the ecological equilibrium of Mexico City. According to Pezzoli (1998: 211), land speculation in Ajusco was initiated in 1974 with the construction of a scenic highway, which—according to several researchers—was "ordered by the then secretary of the highway department so that he could get to and from his residential estate."

In a similar pattern, in recent years high-income gated communities have started to appear around Istanbul's water-basins and in its northern protected forest areas, what were once associated with squatter settlements. Development of these new residences, among other reasons, is motivated around the mayor's grandiose vision for the city's development and erecting a new bridge on the northern part of the city.

In Italy, oversight or encouragement by public officials in construction and development activities is a common sight, especially in the Southern regions, where illegal constructions are attributed to different income groups. These developments range from those on the fertile slopes of Mount Vesuvius to coastal developments, and they are estimated to have risen 30 % in 2003 under the leadership of President Berlusconi and his amnesty laws. In recent years, 600 illegal constructions were discovered in an archeological park in the Sicilian Coast, as the region's mayor, who himself owned one of these residences had allowed their construction in exchange of votes (Sylvers 2004).

Vulnerability due to inadequacy or inefficient application of construction standards and building design, unavailability or disregard of planning, and

---

[25] In studying Latin American cities, Gilbert (1996: 93) argued that "hilly cities are arguably less clearly polarized than flat cities," as high-income and middle-income areas develop in close proximity to *barrios* and *favelas* on steep slopes unsuitable for formal-sector construction. Gilbert (1996: 93) wrote: "Here, every exclusive residential development appears to have its low-income neighbour next door. A functional symbiosis has developed; the urbanización provides work for the maids, shoe menders, laundresses, and the like, and the *barrio* provides cheap labor".

corruption or mismanagement by related officials are experienced everywhere from developing countries to most developed nations. In 2005, a Japanese architect admitted to falsifying building earthquake resistance data on several projects to cut costs and to win contracts. The architect was involved with two hundred structures, including high-rise residential towers, hotels, and temples.[26] This scandal also involved two private building certification firms, which were given authority in 1998 to certify the soundness of new constructions, as part of the government's new policy to deregulate building industry.[27]

Political-decision making combined with poor design and land-use practices have increased the vulnerability of the ecologically hazardous Los Angeles, as well. In his work of disasters in Southern California, Davis (1998) argues that flood, fire, and earthquake tragedies of the region were unnatural and avoidable, and that they occurred as a result of generations long "market-driven urbanization that has transgressed environmental commonsense." In *Ecology of Fear*, Davis describes "historic wildfire corridors turn into view-lot suburbs, wetland liquefaction zones into marinas, and floodplains into industrial districts and housing tracts" (Davis 1998: 9). As urbanization, Davis writes, "relentlessly eroded flood control capacity by paving over watershed and reducing surface absorption, more than 110,000 homes adjacent to the Los Angeles River and Rio Hondo have become vulnerable" (Davis 1998: 36). Construction quality produced other vulnerabilities in the Los Angeles area. Building inspections after the 1994 Northridge earthquake revealed that at least one-third or more of damage in residential buildings were directly related to substandard construction. Huge pre-cast concrete department stores demonstrated similar problems in design and construction, and one expert summarized the situation as a "dangerous combination of inadequacies in building codes and an increasing drive to cut costs by designing for the minimum" (Davis 1998: 44).

In a similar manner, in Florida, investigations after the 1992 Hurricane Andrew found out "major shortcomings in construction techniques and code enforcement" (Mileti 1999: 128). Accordingly, in Southern Dade County, homes built after 1980 in new design trends suffered more damages than pre-1980 constructions. Loss of roof materials, which also let to damage in other buildings and cars, was the most frequently observed type of damage (Mileti 1999). A review of "the county's Board of Rules and Appeals found a number of instances in which changes were made under pressure from builders in the name of construction cost savings," such as the allowing of builders to use staples instead of nails to install roofs (Mileti 1999: 131). Such cases indicate that vulnerability to natural disasters can exist regardless of economic well-fare, creating an imminent danger on urban residents and increasing the need for a variety of vulnerability and risk reduction startegies and actions in urban areas (Fig. 2.7).

---

[26] "Japanese architect falsified earthquake data", in *Architectural Record*, 2006. News Briefs.

[27] "Earth-shaking news", in *Economist* (December 2005): 46.

Fig. 2.7 A home partially destroyed by Super Storm Sandy, Staten Island, New York (Photograph by author, 2012)

2.3.1.3 Risk Reduction Strategies

Similar to the vulnerability of communities, disaster risk reduction actions and management are also affected by social and cultural influences, personal and governmental decision-making, and legal, institutional, and economic constraints. This section will explore risk reduction activities than can be employed by local governments in order to reduce physical vulnerability. These actions can range from land-use planning to building codes and engineering, insurance and economic incentives, and public awareness campaigns; although the focus here will be that on physical planning measures.

As part of the Second Natural Hazard Assessment study in the United States, Olshansky and Kartez (1998) classified actions representing "land use management tools" to guide development in hazard-prone areas. Olshansky and Kartez (1998: 170–174) categorized these tools as:

1. *Building standards*, such as traditional building codes, flood proofing requirements, seismic design standards, and retrofit requirements for existing buildings.
2. *Development regulations* including zoning and subdivision ordinances such as flood-zone regulations, setbacks from faults, steep slopes and coastal erosion areas, and zoning standards for sensitive lands as wetlands, dunes, and hillsides.
3. *Critical and public facilities policies* to move location of public or other important facilities (such as schools, fire stations, hospitals, hazardous materials and utilities) outside of hazard areas in order to discourage development and reduce damages.

4. *Land and property acquisition* in hazardous areas with public funds and using these properties in minimally vulnerable ways. Acquisition of open space, recreation, or undeveloped lands for mitigation; relocation of existing hazard area development and acquisition of development rights.
5. *Taxation and fiscal policies* to provide incentives for people who reduce public costs in hazardous areas by applying regulations for safety, or relocating and reducing density in hazardous areas. Adversely these policies would increase taxes for those who add to the public costs of hazard area development.
6. *Information dissemination* to influence public behavior especially of real estate customers by bringing hazard disclosure requirements for real estate sellers, provide public information such as posting warning signs in high-hazard areas and education of construction professionals.

A number of studies[28] conducted in the United States, between 1979 and 1993, examined local government approaches in the application of these management tools for natural hazards mitigation. According to a summary of the findings of these studies, in highly hazard-prone communities, zoning ordinances and building standards are the most frequently used mitigation tools by local officials in order to regulate private construction in hazard-prone areas. On the other hand, in most cases public officials do not have a comprehensive approach to hazard mitigation, for example by not extending their policies for awareness programs or not using relocation strategies (Olshansky and Kartez 1998: 176–177).

Some of these studies and others, dating from 1979 to 1994, have also explored the factors that influence the adoption of hazard mitigation policies by local governments in the United States. Olshansky and Kartez (1998: 179–187) summarize the results of these studies[29] in two major categories as: controllable and uncontrollable factors. Accordingly, factors controllable by local governments

[28] These were survey studies on local government approaches to hazards. There was a high survey response rate ranging between 75 and 90 %; and types of informants were local planning directors or designees, and local flood coordinators. For more information on floodplain hazards, see "Coping with floods" (Burby and French 1981) and *Flood Plain Land Use Management* (Burby and French 1985). For coastal storms and hurricanes, see *Catastrophic Coastal Storms* (Godschalk et al. 1989). For earthquakes, see "A national assessment of local earthquake mitigation" (Berke et al. 1992), and for multiple natural hazards see *Sharing Environmental Risks* (Burby et al. 1991), and Factors Promoting Comprehensive Local Government Hazards Management (Kartez and Faupel 1995).

[29] These are conclusions derived from the following studies: *The Politics and Economics of Earthquake Hazard Mitigation* (Alesch and Petak 1986); "Hurricane vertical shelter policy" (Berke 1989); "A national assessment of local earthquake mitigation" (Berke et al. 1992); *Flood Plain Land Use Management* (Burby and French 1985); "Mandates, plans and planners" (Dalton and Burby 1994); *Earthquake Mitigation Policy* (Drabek et al. 1983); *Catastrophic Coastal Storms* (Godschalk et al. 1989); *Analysis of Adoption and Implementation of Community Land-use Regulations for Floodplains* (Hutton et al. 1979); *Role of States in Earthquake and Natural Hazard Innovation at the Local Level* (Lambright 1984); Seismic Hazard in the Central United States (Olhansky 1994); and *Preparing for California's Earthquakes: Local Government and Seismic Safety* (Wyner and Mann 1986).

range from recognition of the problem to staff resources, lack of persistent policy advocates, interactions among participants in policy development, and linkage of hazards to other issues such as those that could reinforce the solution of another problem (Gencer 2007, 2008: 286).

Factors that are uncontrollable by local governments include community wealth and resources, "window of opportunity" that opens by local or external disasters, which can increase public awareness and attract federal and state resources, previous hazard experience, lack of "public minded" communities, national regulations and assistance, and the presence of feasible policy solutions. On the other hand, some of the factors that are described to be uncontrollable by local governments do not need to be so. Presence of a feasible policy solution or increasing awareness to create opportunity to integrate mitigation policies into local development plans should be a concern of local governments, which should be proactive rather than waiting for hazards and the subsequent disasters to occur (Gencer 2007, 2008: 286–287).

In 2001, the Institute for Business and Home Safety (IBHS) worked with the American Planning Association (APA) and the American Institute of Certified Planners (AICP) to survey nearly one thousand five hundred municipal-level planners with a questionnaire named "Community Land-Use Evaluation for Natural Hazards."[30] Steinberg and Burby (2002: 22), two leaders of this study, produced a "Growing Safe" plan for communities based on eight fundamental elements:

1. Basics: a general or comprehensive plan and a planning staff;
2. Quality of data: a plan that includes or references factual data and maps;
3. Identification of issues: an explanation of natural hazards and other community issues;
4. Community support: community involvement in preparing the plan;
5. Policies: specific policies addressing hazards;
6. Coordination: consistency with federal, state, regional, and internal community plans;
7. Implementation: goals linked to specific actions;
8. Presentation and organization: a plan that is reasonable, comprehensible and easy to use.

The results of this survey indicated that as an average, communities got a grade of 48 % in this "Growing Safe" report card and that 8 % of the communities scored zero. On four of these elements—plan basics, citizen involvement, consistency and organization—plans on average scored above 50 %, whereas in the remaining issues that addressed natural hazards, they scored equal to or less than 40 %. As important factors affecting their efforts, planners identified the need for "public demand for hazards planning, followed by additional funding, support from elected officials, and technical assistance" (Steinberg and Burby 2002: 23).

[30] Institute for Business and Home Safety (IBHS) (2002).

Additional needs were "better mapping and data; state mandates for planning, additional staff and legislative changes" (Steinberg and Burby 2002: 23). Indeed, Steinberg and Burby concluded that communities,[31] which were located in states that mandated local planning, and which applied safe growth strategies, had higher ratings than the others did, and that state-mandated local comprehensive plans are "the key to better performance" (Steinberg and Burby 2002: 23).

A smaller-scale study founded by the National Science Foundation (NSF) examined whether the quality of local plans changed over an eight-year period from 1991 to 1999 in jurisdictions in Florida and Washington—the two states that scored high grades in the "Growing Safe" project. The principal investigator in this study, Brody (2003) explained that the study examined which hazard mitigation components had changed in the study communities and identified factors that influenced the adoption of new mitigation tools. Plan quality was conceptualized in three components: a strong factual basis, clearly articulated goals, and appropriately directed policies (Brody 2003: 194). Results of the study indicated a significant increase in plan qualities in both states. Accordingly, "plans in Florida showed particular improvements in emergency preparedness such as evacuation and sheltering capabilities. Jurisdictions in Washington strengthened their policies to protect areas subject to flooding through permitted land uses, setbacks, and locating public facilities outside of hazard prone areas" (Brody 2003: 198). As in the IBHS study, findings also suggest that initial quality of plans and legal reform mandates by state authorities had an important effect on the planning communities. The driving force behind this increase in plan quality was different in both states. In Florida, the plan quality "appeared to be driven primarily both by a previously established policy-making momentum and repetitive loss to specific properties" (Brody 2003: 198). In Washington, the planning capacity was influenced most strongly by citizen participation. These results add to the previous findings, which indicated that the factors that impact local governments' integrated mitigation–land use planning processes range from national policy-making, public awareness and involvement, and institutional capacity at the local level.

In the United States (U.S.) federal level, the *Disaster Mitigation Act of 2000* (U.S. Congress 2000), required state and local communities to have an approved mitigation plan in place by November 2004 in order to be eligible for pre- and post- hazard mitigation grant funds; emphasizing the importance of planning before disasters occur. FEMA's (Federal Emergency Management Agency) *How-to-Guide for State and Local Mitigation Planning* provides guidance to local governments, and proposes an inventory assessment for estimating losses from disasters.[32] This assessment requires two major tasks: (1) determining the proportion and the value of buildings, and (2) determining the population located in

---

[31] Among them were statewide Florida, large cities and counties in Nevada, coastal region in North Carolina, statewide Oregon, coastal California, and growth management jurisdictions in Washington State (Steinberg and Burby 2002: 23).

[32] Federal Emergency Management Agency (FEMA) (2001).

hazard areas. The Guide gives the option of extending this inventory to critical facilities, vulnerable populations, major economic elements, high-density residential areas, historic, cultural, and natural resource areas, and other important facilities such as major employers, banks, and gas stations.[33] However, this way of presenting the "detailed inventory" as an option gives way to the preparation of incomplete mitigation plans by local governments with inadequate staff or resources.

As an addition to its *how-to-guide*, FEMA provided requirements on assessing vulnerability in a later document, *Multi-Jurisdictional Mitigation Planning.*[34] FEMA categorizes these requirements in three criteria. Accordingly, mitigation plans should describe vulnerability in terms of: (a) types and numbers of existing and future buildings, infrastructure, and critical facilities located in the hazard area, (b) potential dollar losses to these identified vulnerable losses, and (c) providing a general description of land uses and development trends within the community, so that mitigation options can be considered in future land use decisions.[35] This categorization eliminates the human factor signaling the problematic way disaster mitigation is comprehended in the federal level. On the other hand, despite these shortcomings, these documents can be accepted as an initiation to the acceptance of mitigation as an important part of disaster management, and together with the Act of 2000 provide a base for more comprehensive federal programs and legislations that have been criticized by practitioners and academicians for its patchwork nature.[36]

In the international arena, many local governments have undertaken integrated disaster risk management programs, as multilateral organizations have shifted their focus and assistance from recovery and reconstruction to disaster management. As it was explored in Sect. 2.3.1.1 of this book, in the example of slum upgrading projects supported by financial and technical assistance, integrated disaster risk management programs aim at integrating risk reduction actions into local government services. These actions vary from vulnerability and risk analysis to public awareness and participation, protecting critical infrastructure, and at times larger scale projects aiming at citywide mitigation and disaster risk reduction. Some of the successful and initiating programs range in context. One of them is the Municipal Disaster Mitigation system in Manizales, Colombia based on municipal development and land-use plans, in addition to tax-incentives and voluntary housing insurance scheme.[37] In Manizales, the disaster risk management plan is integrated into the city's development plan and its environmental policy and action plan, and it has been able to bring

---

[33] Federal Emergency Management Agency (FEMA) (2001).

[34] Federal Emergency Management Agency (FEMA) (2006).

[35] Federal Emergency Management Agency (FEMA) (2006): 26–28.

[36] See "Governing Land Use in Hazardous Areas with a Patchwork System" (May and Deyle 1998: 57–82) for legal programs and policies that influence land-use and development in hazard-prone areas.

[37] United Nations Development Programme (UNDP) (2004): 63.

together the local and regional government, the private sector, universities and representatives of community organizations into a participative process.[38] The Municipal Flood Management system in Cologne, Germany is another project deemed successful involving engineering systems, public awareness, emergency management, and integration of Geographical Information Systems (GIS) based flood-risk plans in urban planning (UN/ISDR 2004: 1:130).

In addition to single projects, many regional and international programs support local governments' initiatives in disaster risk reduction. Initiated in 1995 by the Asian Disaster Preparedness Center, the Asian Urban Disaster Mitigation Program (AUDMP) promotes "strategic approaches to urban risk reduction as part of urban development planning processes" (UN/ISDR 2004: 1:134). AUDMP's project activities concentrate on different issues in accordance with local priorities in ten Asian countries, covering activities such as hazard mapping and risk assessment, mitigation planning and implementation, public awareness and education, capacity-building, safer building construction, community-based approaches, and policy, legal, and institutional arrangements (UN/ISDR 2004: 1:135).

Another international program is the "Resilient Communities" project developed by the *International Council for Local Environmental Initiatives* (ICLEI). According to this project, "a resilient community encompasses the acceptance of developing capacities to identify vulnerabilities and activities to reduce them. It employs tools and strategies for hazard reduction and risk management, which include planning measures, urban design features, regulations that are enforced and the investment of resources to protect important assets. It also needs to support institutional and community-based systems for crisis management, response and recovery when necessary" (UN/ISDR 2004: 1:139). As part of this agenda, the *Earthquakes and Megacities Initiative* (EMI) developed a tool known as a *Disaster Management Master Plan* (DMMP). Wenzel and Bendimerad (2002: 119–124) of the EMI explain that a DMMP consists of five components: (1) disaster assessment, (2) disaster preparedness, (3) disaster response and relief, (4) disaster mitigation, and (5) know-how and expertise acquisition. They argue that DMMP is "a rational and efficient approach to build local capacity because it fits conventional local government operating framework," which is "driven by similar plans in areas such as urban development, land-use planning, capital planning and public safety" (Wenzel and Bendimerad 2002: 121). In recent years, with the initiation and support of international agencies, a number of hazard-prone metropolitan cities, such as Mumbai[39] and Istanbul, have prepared Disaster Management Master Plans, emphasizing institutional and legal changes that may pave the way to the possible employment of these intensive studies.

Another program, the *Mayor's Task Force on Climate Change, Disaster Risk and the Urban Poor*, which was launched at the Mayor's Summit in Copenhagen

---

[38] Satterthwaite (2011): 16–17.

[39] See "Disaster management plan for the State of Maharashtra, India" (Vatsa and Joseph 2003) for the Mumbai case.

in 2009, identifies good practice examples and propose policy and investment programs to improve the resilience of the urban poor to disaster risks and climate change. As part of a global study carried out by the World Bank as part of the Mayor's Task Force work program, the following actions are recommended to build resilience of the urban poor: (a) assessing risk at the city and community level, (b) integrating climate change and disaster risk reduction policies for the poor in urban planning and management, (c) building institutional capacity to deliver basic services and reduce vulnerability to climate and disaster risk, (d) bridging communities and local governments to work together on local solutions, and (e) opening new finance opportunities for cities to address climate change adaptation and disaster risk reduction (WB 2011).

And most importantly the UNISDR's 2010–2015 global campaign proposes a 10 step checklist for *Making Cities Resilient*:

1. Put in place organization and coordination to understand and reduce disaster risk, based on participation of citizen groups and civil society. Build local alliances. Ensure that all departments understand their role in disaster risk reduction and preparedness.
2. Assign a budget for disaster risk reduction and provide incentives for home-owners, low income communities, businesses and the public sector to invest in reducing the risks they face.
3. Maintain up to date data on hazards and vulnerabilities. Prepare risk assessments and use these as the basis for urban development plans and decisions, ensure that this information and the plans for your city's resilience are readily available to the public and fully discussed with them.
4. Invest in and maintain critical infrastructure that reduces risk, such as flood drainage, adjusted where needed to cope with climate change.
5. Assess the safety of all schools and health facilities and upgrade these as necessary.
6. Apply and enforce realistic, risk compliant building regulation and land use planning principles. Identify safe land for low income citizens and upgrade informal settlements, wherever feasible.
7. Ensure that education programmes and training on disaster risk reduction are in place in schools and local communities.
8. Protect ecosysystems and natural buffers to mitigate floods, storm surges and other hazards to which your city may be vulnerable. Adapt to climate change by building on good risk reduction practices.
9. Install early warning systems and emergency management capacities in your city and hold regular public preparedness drills.
10. After any disaster, ensure that the needs of the affected population are placed at the centre of reconstruction, with support for them and their community organizations to design and help implement responses, including rebuilding homes and livelihoods (UN 2012).

2.4 Conclusion

Disasters and development have an interlinked and multifaceted relationship. They can mutually have a negative effect on each other. On the other hand, sustainable development can also help reduce disaster risks. Today, research on this complex relation is more essential than at any other time in history, as worldwide statistics indicate an increasing number of disasters as recent patterns, and as climate change is expected to increase the intensity and severity of hazards in urban areas.

Disaster statistics indicate the increasing impacts of disasters (with the exception of mortalities) within the last decades, and they reveal a general pattern in relation to geographical location and development. For instance, within the last three decades, Asia had the highest number of geophysical, hydrological and meteorological disasters, Africa experienced the highest number of biological disasters and droughts and Europe had the highest number of climatological disasters. These results indicate the global spread of disaster impacts, the existence of a variety of vulnerabilities in relation to development levels, and a need for different types of disaster risk management.

Vulnerability to natural disasters is expected to be increasingly affected by the global force of urbanization. Urbanization, together with other interlinked forces, can either generate or increase intensity of hazards (such as with climate change and land degradation), as well as having the potential to increase vulnerability to hazards (such as with globalization).

This chapter has described two images that represent disaster vulnerability of urban populations ranging from those in low-income countries to middle- and higher-income ones. It has shown that there is a strong tie between vulnerability and urban poverty, and that an understanding of urban poverty encompassing both economic and non-economic factors provides insight to disaster vulnerability in urban areas, such as in informal settlements and slums. In order to understand the full extent of the sources of urban poverty and vulnerability, it is also essential to gain an overall coherence of rural–urban linkages and to promote mutual policies such as those for land tenure, appropriate land allocation or interregional transport and infrastructure.

This chapter has also argued that, on the other hand, it is necessary to stress that *vulnerability* is not identical with *poverty*; and that "not all poor people are vulnerable to disasters, and some people who are not poor are also vulnerable" (Bankoff 2003: 19). In some instances, communities move towards new design and construction schemes with untrained professionals or insufficient inspections; or in other cases, they disregard spatial planning schemes leading to the vulnerability of these communities.

The study of various risk reduction programs in urban areas confirmed that there is no one solution to disaster mitigation and different strategies need to be applied to the needs of diverse communities. However, it also revealed the persistent need of conditions such as "good urban governance",[40] planning, building, and economic measures for successful risk reduction strategies. Public awareness, empowerment, and participation of urban residents are key elements to reducing vulnerability in urban areas, providing not only motivation of residents but also success in implementation with real space–time input. Coordination of involved organizations, knowledge sharing, and data gathering are other essential components. A willing and proactive local government with financial and technical resources (as part of good urban governance) is one of the foremost requirements to be able to implement integrated risk reduction practices.

Physical planning, construction and building design standards are essential elements in urban disaster risk management. However, as much as adequate zoning, building regulations, and legal tools are necessary, they can sometimes be too rigid and expensive for urban residents to employ, leading the way to an untrained informal construction sector and settlements. It was observed that the failure to analyze costs of imposing certain zoning regulations in advance "can easily imply that well-intended regulation will end up hurting the poor" (Deininger 2003: 176). Evidence has shown "the inverse relationship between informality and the imposition of regulations" in many developing countries (Deininger 2003: 176). The measures to meet strict land-use and building regulation are found "too expensive or bureaucratically cumbersome" for many, "pushing more and more housing and settlements outside the regulations".[41] Again, although some local governments develop master plans to regulate urban development and expansion, lack of consultation with cities' residents and interest groups lead to poor results in their application.

These problems stress the significance of good urban governance in bringing together different groups to input for decisions concerning the future of the city. Such a multi-dimensional planning process can provide the way to reducing disaster risk while producing a sustainable urban development, where "environmental quality, economic growth and social justice coexist" (Beauregard 2003). The next chapter will examine how the lack of such a planning and development process has resulted in the exposure and vulnerability of the hazard-prone city of Istanbul and its residents.

---

[40] According to the UN-Habitat's Governance Campaign, principles of "good urban governance" is characterized by *sustainability* in all dimensions of urban development, *subsidiarity* of authority and resources to the closest appropriate level, *equity* of access to decision-making processes and the basic necessities of urban life, *efficiency* in the delivery of public services and in promoting local economic development, *transparency and accountability* of decision-makers and all stakeholders, *civic engagement and citizenship*, and *security* of individuals and their living environment. United Nations Human Settlements Program (UN-Habitat), 2004: Principles of Good Urban Governance; at: http://www.unhabitat.org (2006):3–7.

[41] Satterthwaite (2011): 19.

References

Ames B, Brown W, Devarajan S, Izquierdo A (2002) Macroeconomic issues. In: Klugman J (ed): A sourcebook for poverty reduction strategies, vol 2. World Bank, Washington, pp 3–27

Alesch DJ, Petak WJ (1986) The politics and economics of earthquake hazard mitigation. Institute of Behavioral Science, University of Colorado, Boulder

Baharoglu D, Kessides C (2002) Urban poverty. In: Klugman J (ed) A sourcebook for poverty reduction strategies, vol 2. World Bank, Washington, pp 123–159

Baker JL (2006a) Integrated urban upgrading for the poor: the experience of Ribeira Azul, Salvador, Brazil. World Bank policy research working paper 3861. http://www-wds.worldbank.org/external/default/WDSContentServer/IW3P/IB/2006/02/22/000016406_20060222115502/Rendered/PDF/wps3861.pdf

Bankoff G (2003) Vulnerability as a measure of change in society. Mass Emerg Disasters 21(2):5–30

Beauregard R (2003) Democracy, storytelling, and the sustainable city. In: Eckstein B, Thogmorton JA (eds) Story and sustainability. MIT Press, Cambridge, pp 65–77. Quoted in Fainstein S (2005) Planning theory and the city. J Plann Educ Res 25:121–130, 126

Berke PR (1989) Hurricane vertical shelter policy: the experience of Florida and Texas. Coast Manag 17:193–217

Berke PR, Beatley T (1992) A national assessment of local earthquake mitigation: implications for planning and public policy. Earthq Spectra 8(1):1–15

Bigio AG (2003) Cities and Climate Change In: Kreimer A, Arnold M, Carlin A (eds) Building safer cities. Disaster risk management series no: 3, World Bank, Washington, pp 91–99

Bolt B (2004) Earthquakes, 5th edn. W. H. Freeman and Company, New York

Brody SD (2003) Are we learning to make better plans? A longitudinal analysis of plan quality associated with natural hazards. J Plann Educ Res 23:191–201

Burby RJ, French SP (1981) Coping with floods: the land use management paradox. J Am Plann Assoc 47:289–300

Burby R. French SP (1985) Flood plain land use management: a national assessment. In: Cigler B, Kaiser EJ, Moreau DH, Stiftel B (eds) Westview Press, Boulder

Burby RJ, Cigler BA, French SP, Kaiser EJ, Kartez J, Roenigk D, Weist D, Whittington D (1991) Sharing environmental Risks: How to control government's losses in natural disasters. Westview Press, Boulder

Cutter S (ed) (2001) American hazardscapes: the regionalization of hazards and disasters. Joseph Henry Press, Washington

Dalton L, Burby RJ (1994) "Mandates, plans, and planners: building local commitment to development management. J Am Plann Assoc 60 (Autumn):444–462

Davis M (1998) Ecology of fear: Los Angeles and the Imagination of Disaster. Metropolitan Books/Henry Holt and Company, New York

Deininger K (2003) Land policies for growth and poverty reduction. A orld Bank policy research report. World Bank/Oxford University Press, Washington

Dodman D, Hardoy J, Sattherwaite D (2009) Urban development and intensive and 1154 extensive risk. Background paper to the 2009 global assessment report on disaster risk 1155 reduction, Geneva. http://www.preventionweb.net/1156 english/hyogo/gar/background-papers/

Drabek TE, Mushkatel AH, Kilijanek TS (1983) Earthquake mitigation policy: the experience of two states. program on environment and behavior. Monograph 31 institute of behavioral science, University of Colorado, Boulder

Earthquake Engineering Research Institute (EERI) (2003) The boumerdes, algeria, earthquake May 21, 2003: An EERI learning from earthquakes reconnaissance report

Eswaran H, Berberoglu S, Cangir C, Boyraz D, Zucca C, Ozevren E, Yazici E, Zdruli P, Dingil M, Donmez C, Akca E, Celik I, Watanabe T, Koca YK, Montanarella L, Cherlet M, Kapur S (2011) The anthroscape approach in sustainable land use. In Kapur S, Eswaran H, Blum WEH (eds) Sustainable land management: learning from the past for the future. Springer, Berlin, pp 1–50

Federal Emergency Management Agency (FEMA) (2001) State and local mitigation planning how-to-guide: understanding your risks.http://www.fema.org (2006)

Federal Emergency Management Agency (FEMA) (2006) Multi-jurisdictional mitigation planning: state and local mitigation planning how-to-guide 368–368. http://www.fema.org

Gencer EA (2007) The interplay between natural disasters, vulnerability, and sustainable development. Ph.D. dissertation, Columbia University, New York

Gencer EA (2008) Natural disasters, vulnerability, and sustainable development: examining the interplay, global trends, and local practices in Istanbul. VDM Verlag, Saarbrucken

Gilbert A (1996) Land, housing, and infrastructure in Latin America's major cities. In: Gilbert A (ed) The mega-city in latin America. The United Nations University Press, Tokyo, pp 73–109

Godschalk DR, Brower DJ, Beatley T (1989) Catastrophic coastal storms: hazard mitigation and development management. Duke University Press, Durham

Hardoy J, Satterthwaite D (1993) Housing policies: a review of changing government attitudes and responses to city housing problems in the third world. In: Cheema GS (ed) Urban management: policies and innovations in developing countries, 111–159 Praeger Publishers, Tokyo. United Nations University, Honolulu. Hawaii East–West Center, Westport

Hutton JR, Mileti DS, Lord WB, Sorensen JH, Waterstone M (1979) Analysis of adoption and implementation of community land-use regulations for floodplains. Woodward-Clyde Consultants, San Francisco

Institute for Business and Home Safety (IBHS) (2002) Are we planning safer communities? Results of a national survey of community planners and natural Disasters. http://www.ibhs.org/publications

Intergovernmental Panel on Climate Change (IPCC) (2001) Climate change 2001: impacts, adaptation and vulnerability summary for policy makers and technical summary for working group II report, IPCC, Geneva. Quoted in UNDP (2004) Reducing disaster risk: a challenge for development, 69. http://www.undp.org/bcpr

Intergovernmental Panel on Climate Change (IPCC) (2012) Summary for policymakers. In: Fields CB, Barros V, Stocker TF, Qin D, Dokken DJ, Ebi KL, Mastrandrea MD, Mach KJ, Plattner GK, Allen SK, Tignor M, Midgley PM (eds) Managing the risks of extreme events and disasters to advance climate change adaptation: a special report of the working groups I and II of the intergovernmental panel on climate change. Cambridge University Press, Cambridge, pp 1–19

Kante B (2005) The urban–rural environment linkages: integrating the brown and green environment agendas. in UN-Habitat, Urban-Rural Linkages Approach to Sustainable Development, 244–255 (UN-Habitat)

Kartez J, Faupel C (1995) Factors promoting comprehensive local government hazards management. In: From the mountains to the sea—developing local capabilities. Proceedings of the nineteenth annual conference of the association of state floodplain managers. Natural hazards research and applications information center special publication 31. Institute of Behavioral Science, University of Colorado, Portland

Lambright WH (1984) Role of states in earthquake and natural hazard innovation at the local level: a decision making study. National Technical Information Service, Washington

Laquian A (2005) Beyond metropolis. Woodrow Wilson Center Press and Baltimore: John Hopkins University Press, Washington

Leontidou L (1990) The mediterranean city in transition: social change and urban development. Cambridge University Press, Cambridge

Masselos J (1995) Postmodern Bombay: fractured discourses. In: Watson S, Gibson K (eds) Postmodern cities and spaces. Blackwell, Oxford, pp 199–215

May PJ, Deyle RE (1998) Governing land use in hazardous areas with a patchwork system. In: Burby RJ (ed) Cooperating with nature: confronting natural hazards with land-use planning for sustainable communities. Joseph Henry Press, Washington, pp 57–82

McCarthy JOJ, Canziani F, Leary NA, Dokken DJ, White KS (eds) (2001) Climate change 2001: impacts, adaptation, and vulnerability. contribution of working group II to the third assessment report of the intergovernmental panel on climate change. Cambridge University Press, Cambridge

McGranahan G, Jacobi P, Songsore J, Surjadi C, Kjellén M (2001) The citizens at risk: from urban sanitation to sustainable cities. Earthscan, London

Meli R (1993) Earthquake resistant design and construction of reinforced concrete buildings: the practice in Mexico after 1985. In: Proceedings of the 1993 structures congress: structural engineering in natural hazard mitigation, vol 1. ASCE, New York pp 229–234. Quoted in Meli R, Alcocer SM (2004) Implementation of structural earthquake-disaster mitigation programs in developing countries. Nat Hazards Rev 5(1):31

Mileti DS (1999) Disasters by design: a reassessment of natural hazards in the United States. Joseph Henry Press, Washington

Meli R, Alcocer SM (2004) Implementation of structural earthquake-disaster mitigation programs in developing countries. Nat Hazards Rev 5(1)29–39

Olhansky RB (1994) Seismic hazard in the central United States: the role of the states. Professional Paper 1538-G. Geological Survey, Reston

Olshansky RB, Kartez JD (1998) Managing land use to build resilience. In: Burby RJ (ed) Cooperating with nature: confronting natural hazards with land-use planning for sustainable communities. Joseph Henry Press, Washington, pp 167–201

Oliver-Smith A (1999) Lima, Peru: underdevelopment and vulnerability to hazards in the city of the kings. In: Mitchell JK (ed) Crucibles of hazard: mega-cities and disasters in transition. United Nations University Press, Tokyo

Pelling M (2003) The vulnerability of cities: natural disasters and social resilience. Earthscan, London

Perlman JE (1993) Mega-cities: global urbanization and Innovation. In: Cheema GS (ed) Urban management: policies and innovations in developing countries. Praeger Publishers, United Nations University, East-West Center, Westport, pp 19–50

Pezzoli K (1998) Human settlements and planning for ecological sustainability: the case of Mexico City. The MIT Press, Cambridge

Puente S (1999) Social vulnerability to disasters in Mexico City: an assessment method. In: Mitchell JK (ed) Crucibles of hazard: mega-cities and disasters in transition. United Nations University Press, Tokyo, pp 295–335

Sachs J (2005) The end of poverty: economic possibilities for our time. The Penguin Press, New York

Sanderson D (2000) Cities, disasters and livelihoods. Environ Urbanization 12(2):93–102

Satterthwaite D (2005) The scale and nature of urban change worldwide: 1950–2000 and its underpinnings. Human settlement discussion paper series. International Institute for Environment and Development (IIED), London

Satterthwaite D (2011) What role for low-income communities in urban areas in disaster risk reduction?" Background paper to the 2011 global assessment report on disaster risk reduction, Geneva http://www.preventionweb.net/english/hyogo/gar/backgroundpapers/ pp 16–17

Sattherwaite D, Cecilia T (2002) Seeking an understanding of poverty that recognizes rural-urban linkages. In: Rakoli C, Lloyd-Jones T (eds) Urban livelihoods: a people centered approach, Earthscan, London, pp 52–70

Steinberg M, Burby RJ (2002) Growing safe. Planning 68(4):2–23

Spaliviero M (2006) Integrating slum upgrading and vulnerability reduction in Mozambique. Open House Int 31(1):106–115

Sylvers E (2004) Build illegally on Italy's land and amnesty law will come. New York Times

United Nations (UN) (2005) Report of the world conference on disaster reduction. A/CONF. 206/6, Kobe, Hyogo

United Nations (UN) (2010) Local governments and disaster risk reduction: good practices and lessons learned. UNISDR, Geneva

United Nations (UN) (2012) How to make cities more resilient: a handbook for local government leaders. UNISDR, Geneva

United Nations (UN) Millenium Project (2005) A home in the city. Task force on improving the lives of slum dwellers. Earthscan, London

United Nations Development Programme (UNDP) (2004) Reducing disaster risk: a challenge for developmenthttp://www.undp.org/content/dam/undp/library/crisis%20prevention/disaster/asia_pacific/Reducing%20Disaster%20risk%20a%20Challenge%20for%20development.pdf, p 61

United Nations Human Settlements Programme (UN-Habitat) (2003) The challenge of slums: global report on human settlements 2003. Earthscan, London

United Nations Human Settlements Programme (UN-Habitat) (2009) Planning sustainable cities: global report on human settlements 2009. Earthscan, London

United Nations Human Settlements Programme (UN-Habitat) (2011) Global report on human settlements: cities and climate change. Earthscan, London

United Nations International Strategy for Disaster Reduction (UN/ISDR) (2004) Living with risk: a global review of disaster reduction initiatives, 2 vols. United Nations, Geneva

United Nations International Strategy for Disaster Reduction (UN/ISDR) (2009) UNISDR terminology on disaster risk reduction, UN/ISDR, Geneva

United Nations Population Division (UNPD) (2002) World urbanization prospects: the 2001 revision, data tables and highlights. New York. http://www.un.org/esa/population. Quoted in UN-Habitat: The Challenge of Slums: Global Report on Human Settlements, 26 (London and Sterling, VA: Earthscan)

United States (US) Congress (2000) Disaster mitigation act of 2000. Public Law 106–390. 106th Congress. 30 Oct 2000

Vatsa KS, Joseph J (2003) Disaster management plan for the state of Maharashtra, India: evolutionary process. Nat Hazards Obs 206–212

Wenzel F, Bendimerad F (2002) Earthquakes and megacities. In: Beer T, Ismail-Zadeh A (eds) Risk science and sustainability: science for reduction of risk and sustainable development of society, NATO science series. Kluwer/NATO Scientific Affairs Division, Dordrecht, pp 111–125

Westgate K (1981) Land-use planning, vulnerability and the low-income dwelling. In: Davis I (ed) Disasters and the small dwelling. Pergamon Press, Oxford, pp 27–31

Wines M (2005) Zimbabwe police resume drive to raze slums. New York Times

Wisner B, Blaikie P, Cannon T, Davis I (2004) At risk: natural hazards, people's vulnerability, and disasters. 2nd edn. Routledge, London

World Bank (WB), the Global Facility for Disaster Reduction and Recovery (GFDRR) (2010) Disaster risk management in Latin America and the Caribbean region: GFDRR country notes

World Bank (WB) (2001) World development report: attacking poverty. Oxford University Press, Washington

World Bank (WB) (2003) World Development report. Sustainable development in a dynamic world: transforming institutions, growth, and quality of life. Oxford University Press, Washington

World Bank (WB) (2010) Natural hazards, unnatural disasters: the economics of effective prevention. The International Bank for Reconstruction and Development, Washington

World Bank (WB) (2011) Climate change, disaster risk, and the urban poor. International Bank for Reconstruction and Development (IBRD)/World Bank (WB), Washington

Wyner AJ, Mann DE 919860 Preparing for California's Earthquakes: Local Government and Seismic Safety Institute for Governmental Studies, University of California, Berkeley

Chapter 3
The Making of a Disaster: Earthquake Hazard and Urban Development in Istanbul

The problem is that the damaged buildings are without construction licenses and are at odds with development plans. Out of 15,000 buildings, 2,000 of them were built illegally after 1980 amnesty laws. There are an additional 2,000 buildings that were pardoned by amnesty laws.

B. Yıldırım (Civil Engineer, Zeytinburnu Municipality).
[Author's translation].

Istanbul, the largest and the most populated city in Turkey, lies at the crossroad between two continents. Located on the Strait of Bosphorus, Istanbul is a bridge between Europe and Asia, and through the Sea of Marmara, it links civilizations on the Black Sea with those of the Aegean and the Mediterranean Seas.

With this unique location and cultural background, Istanbul is an important metropolis. Today, home to more than ten million people, the mega-city of Istanbul serves as the commercial and the cultural center, as well as the heart of the Turkish economy. The natural setting that has created the potency of Istanbul's urban environment is, however, also a character that threatens it. Located in an active earthquake zone, Istanbul's history has been interrupted many times by earthquakes; and today history may repeat itself as scientists predict that in the near future, the city will experience a major earthquake.

This chapter examines the links between urban development and vulnerability in the metropolitan city of Istanbul. The chapter starts by examining Istanbul's current earthquake hazard exposure and risk. With a historical analysis, it then analyzes Istanbul's urban development and planning practices pertaining to the city's current exposure and vulnerability. The chapter ends by providing a summary of the results of these studies, and bringing out lessons learned to achieve a disaster resilient and a more sustainable urban development in the metropolitan city of Istanbul.

E. A. Gencer, *The Interplay Between Urban Development, Vulnerability, and Risk Management*, Mediterranean Studies 7,
DOI: 10.1007/978-3-642-29470-9_3, © The Author(s) 2013

3.1 Earthquake Hazard and Risk in the Istanbul Metropolitan Area

This section investigates the current earthquake hazard exposure and risk in the Istanbul Metropolitan Area. It starts by studying the tectonic[1] setting of Istanbul, and then it examines types of damages that occurred in the 1999 Marmara earthquakes, which increased the earthquake stress in Istanbul. This section also reviews hazard assessment and vulnerability analysis for the Istanbul region, undertaken by a consortium of the Istanbul Metropolitan Municipality, identifying the potential risk for the metropolitan city.

3.1.1 Tectonic Setting and the 1999 Marmara Earthquakes

Turkey's unique location between two continents and three major *plates*[2]—African, Eurasian and Arabian—has caused major seismic events over the centuries. As the African and the Arabian plates move north towards the Eurasian Plate, the minor Anatolian plate is caused to move westward, resulting in a *strike-slip fault*,[3] known as *The North Anatolian Fault*, which is similar in length and movement to the San Andreas Fault in California (BU 2002: 87).

The North Anatolian Fault lies from the Eastern Anatolian Region of Turkey to the western edge of the country, passing through the Sea of Marmara. Istanbul's seismic risk results from its location along this sea, under which tectonic plates move on one of the most active geologic boundaries in the world.

In recent years, an east-to-west progression of earthquakes along the North Anatolian Fault has increased Istanbul's earthquake risk. The first of the most recent major earthquakes, which are referred to as the Marmara earthquakes, occurred on 17 August 1999. The epicentre of the magnitude 7.4 earthquake was at Gölcük, south of Izmit, an industrial city that is located east of Istanbul. On 12 November 1999, a second earthquake with a magnitude 7.2 occurred in the region, near the town of Düzce (Fig. 3.1).

---

[1] Tectonic is the geology about the forces that produce movement and deformation of the earth's crust.

[2] According to *plate tectonics* theory, "the Earth consists of large, mobile oceanic and continental regions of solid rocks, called *plates*, floating on softer rock. Plates are in motion and interact with one another through collisions, or slide along or over or under one another" (Bolt 2004: 30).

[3] In strike-strip faults relative displacement of rocks is purely horizontal (Bolt 2004: 357). In these faults, plates moving past each other horizontally lock together until tension builds to a release point. In the process, stress is released and placed on the neighbouring segments of the fault. These neighbouring segments are then more likely to rupture, resulting in progressive earthquakes along the fault (CU 2002: 123–124).

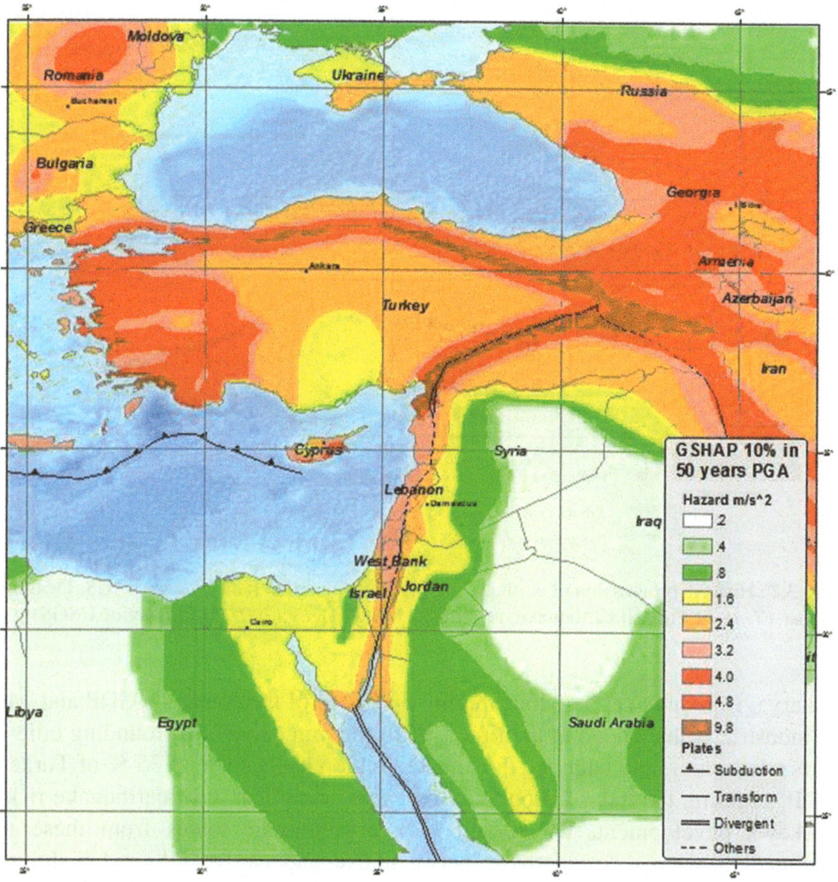

Fig. 3.1 Seismic hazard map of Turkey: peak ground acceleration (m/s2) with 10 % probability of exceedance in 50 Years [*Source* United States Geological Survey (USGS), earthquake hazard program; at: http://earthquake.usgs.gov/regional/world/turkey/gshap.php (2007) (*Courtesy* of USGS)]

The Marmara earthquakes were among the strongest earthquakes in Turkey, and they are considered to be the worst natural disaster in the Mediterranean region between 1975 and 2001 (Brauch 2003). The losses from the two earthquakes were devastating: around 18,000 people lost their lives and 50,000 people were injured. In the two earthquakes, more than 300,000 housing units and 46,000 business premises were damaged, and 320,000 people lost their jobs.[4] The extensive geographical area affected by the earthquake is considered "the industrial heartland of Turkey," with the most severely affected four cities (Kocaeli,

4 Bibbee et al. (2000): 1.

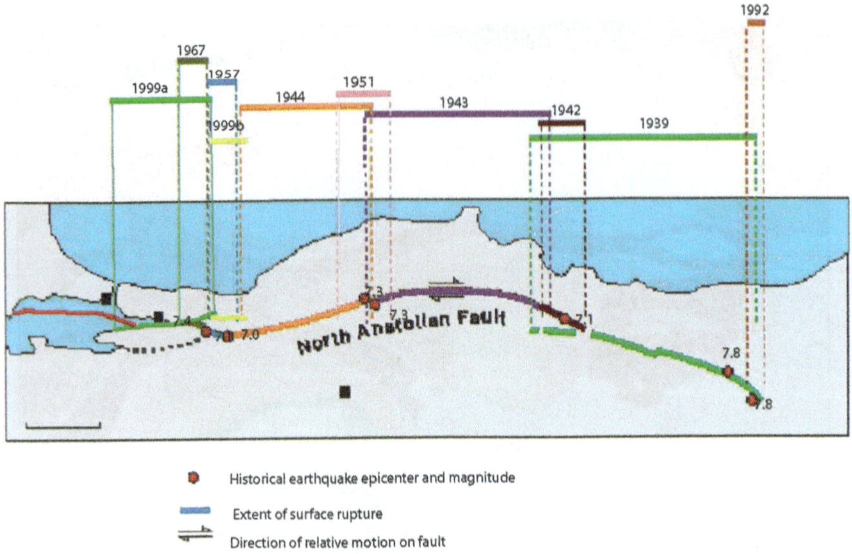

Fig. 3.2 Historic progression of earthquakes on north Anatolian fault [*Source* USGS, location of august 17, 1999 Turkish earthquake; at: http://www.usgs.gov (2007) (*Courtesy* of USGS)]

Sakarya, Bolu and Yalova) contributing over 7 % of the country's GDP and 14 % of industrial value added at the time.[5] With the immediately surrounding cities of Bursa, Eskişehir, and Istanbul, the affected region had a share of 35 % of Turkey's GDP, stressing the risk of economic losses and significance of earthquake risk to Turkey's development. Direct and indirect economic losses from these two earthquakes were reported to be between five and twelve billion US dollars,[6] resulting in the world's highest relief cost between 1992 and 2003 (Fig. 3.2).[7]

The immense damage of the Izmit earthquake pointed to the vulnerability of structures in Turkey. Post-earthquake inspections revealed that fault rupture, ground shaking, and soil liquefaction had caused structural damage that was intensified by poor construction quality.[8] In addition to problems associated with construction techniques and materials, poor planning decision had allowed construction on liquefiable soils or directly over the fault line.

---

[5] Bibbee et al. (2000): 35.

[6] The first number indicates the direct and the indirect losses that have been systematically compiled and reported to date in 2003 World Bank figures. The second numbers by OECD (Organisation for Economic Cooperation and Development) are suggested numbers reported by TUSIAD (Turkish Industrialist's and Businessmen's Association) as well as the numbers from Turkey's State Planning Organization (DPT).

[7] EQE (1999): 37.

[8] EQE (1999): 7.

A major problem was liquefaction.[9] Large-scale urbanizations were permitted on liquefiable soil, resulting in inadequate foundation systems and major loss of lives, especially in the town of Adapazarı, where majority of the buildings that had collapsed were less than twenty years old (USGS 2000). A second problem was fault rupture. Many buildings were destroyed as a result of having been built directly over, or immediately adjacent to the ruptured segment of the North Anatolian Fault. Entire villages and developments including Turkey's largest naval base in Gölcük were destroyed in this manner (USGS 2000).[10]

Most of the severely damaged or totally collapsed buildings were four to eight stories in height, and they were relatively new; in many cases, they were recently completed reinforced concrete frame buildings with masonry infill (USGS 2000: 41). Post-earthquake investigations report that damage to these buildings,[11] was attributed to foundation failures, soft stories (mostly used for commercial purposes) with no shear walls, strong beams and weak columns, lack of column confinement eliminating ductility,[12] and poor detailing practices; all corresponding to substandard construction practices and lack of enforcement of building codes.[13] In most cases, concrete quality was very poor and unacceptably weak; and the presence of seashells in concrete suggested the use of beach sand, which was particularly observed in the debris from the destruction of large apartment buildings for lower-middle class summer housing in the town of Yalova.

3.1.2 *Earthquake Hazard Assessment and Vulnerability Analysis*

Following the Marmara earthquake disasters and the heightened possibility of an earthquake in the Marmara Sea (due to the transfer of stress released by the Izmit earthquake), governmental, international, and academic organizations started working on depicting risk in the Istanbul metropolitan area. As an initial step, earthquake probability[14] calculations were performed for the Marmara fault. In

[9] Liquefaction is "the process by which sandy wet earth materials become fluid like when shaken by earthquakes" (USGS 2000: 3).

[10] On the other hand, an earthquake reconnaissance team member reports that "an individual complex being constructed 100 ft from the fault had very well confined columns with damage limited to spalling and large residual displacements" (Bruneau 1999), pointing to the fact that structurally sound buildings could withstand the destruction regardless of their locations.

[11] In rare cases where steel construction were used (mostly in industrial buildings), damages were attributed to the failure of anchor bolts and structural instability (Bruneau 1999).

[12] Ductility is "the property of a material to deform without catastrophic loss of strength" (USGS 2000: 3).

[13] Bruneau (1999).

[14] Hazard assessments are calculated using several hazard properties such as event magnitude, frequency, speed of onset, time of onset, event duration, temporal spacing (periodicity) and

2000, Tom Parsons and his colleagues announced their calculations, and predicted the probability[15] of an earthquake of M 7 or greater occurring near Istanbul within the next decade to be 32 ± 12 %, and in the next 30 years 62 ± 15 % (Parsons et al. 2000: 1). Following this forecast, deterministic earthquake hazard assessments were undertaken in joint studies of the *Japan International Cooperation Agency* (JICA) and the *Istanbul Metropolitan Municipality* (IMM) (JICA/IMM/PCI/OYO 2002); and of *Boğaziçi University* (2002: 75). These studies concluded that the occurrence of a worst-case scenario earthquake with M 7.5 is considered "highly probable in the next 70 years" (Erdik et al. 2003: 17). Following this diagnosis and the report of the newly founded National Earthquake Council, the Istanbul Metropolitan Municipality organized a consortium to provide risk assessment and vulnerability studies for an Earthquake Master Plan for Istanbul.

The main hazard of earthquakes is caused by ground-shaking, which depends on the combination of various factors, such as the magnitude of the earthquake, the distance from the rupture, and local geological conditions (Smith 2001: 130). According to the site dependent deterministic intensity distribution maps presented in the *Istanbul Earthquake Master Plan* (BU/ITU/METU/YTU 2003), overall conditions lead to a likelihood of strong ground motion on the southern part of Istanbul on the European side, along a portion of the coastline on the Asian side, and the Princes' Islands[16] on the Marmara Sea. The northern areas of the city, due to their relatively large distance from the fault and more stable soils, are expected to have lower site-specific intensities resulting from the occurrence of a scenario earthquake (Erdik et al. 2003). Unfortunately, the hazard exposed areas are also the areas where a large portion of Istanbul's development has taken place, which heightens the potential risk in the city (see Figs. 3.3 and 3.4).

Indeed, risk calculations performed for the predicted scenario earthquake present a bleak picture for Istanbul. Due to deficiencies in design, quality of concrete as a construction material, and construction practices, the majority of the reinforced concrete building stock in Istanbul falls into an average vulnerability class C of the 1998 European Macro-Seismic Scale (EMS). Therefore, intensity based vulnerability calculations for the Istanbul building inventory in 2000 expects more than 40,000 buildings (or about 5.5 % of the total building stock) to be damaged beyond repair, 77,000 buildings (10.5 %) to be substantially-to-heavily

---

(Footnote 14 continued)
spatial dispersion of the event. There are two main approaches that are used in assessing an earthquake hazard: probabilistic and deterministic methods. Cutter (2001: 24) explains that "[t]he *probabilistic approach* attempts to describe the integrated effects from all possible faults at an individual site." *Deterministic approach*, on the other hand, specifies a magnitude or level of ground shaking mostly for a single fault. This approach commonly represents a "worst case scenario," or the maximum risk the location and its residents are exposed to (Cutter 2001: 24).

[15] In 2004, Parsons (2004) of the USGS reassessed the hazard with improved Marmara Sea faulting and a new historical earthquake catalogue and recalculated that the 30-year probability of an earthquake at Istanbul is 41 ± 14 %.

[16] See Appendix for a map highlighting all mentioned Istanbul locations in this book.

Fig. 3.3 Istanbul: space image [*Source* National Aeronautics and Space Administration (NASA); at: http://veimages.gsfc.nasa.gov//16812/ISS008-E-21752_lrg.jpg(2004) (*Courtesy* of NASA)]

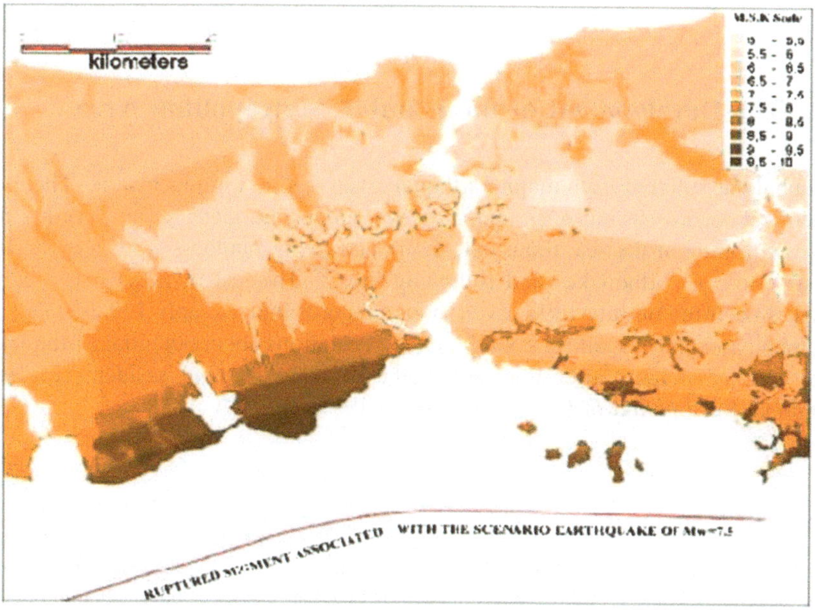

Figure 20. Site-specific intensities in Istanbul that would result from occurrence of the scenario earthquake.

Fig. 3.4 Distribution of intensities after a $M_w = 7.5$ earthquake [*Source* Erdik et al. (2001): 20]

damaged, and about 200,000 buildings (27 %) to be moderately damaged in a scenario earthquake (BU 2002: 243). Casualty calculations, based on both night and daytime population, and that were measured using HAZUS99[17] methodology and estimating a direct relationship between the structural damage and the number of casualties (Erdik et al. 2003: 19), indicate that based on this building vulnerability, casualty levels in a scenario earthquake may vary between thirty to forty thousand people (BU 2002: 243). In addition to the potency of casualty, expected structural damages will also cause loss of settlement habitability. Analysis predicts about 600,000 households to be in need of shelters following a scenario earthquake in Istanbul, without adding the potential short-term needs of residents in moderately damaged houses (BU 2002: 213, 243).

In addition to buildings, other engineered urban structures, infrastructures, and lifelines are also susceptible to the effects of earthquakes. However, physical exposure and casualties represent a portion of what Istanbul can be faced with in case of a worst-case scenario earthquake. According to this assessment in 2002, monetary losses from building damages in a scenario earthquake are estimated to be in the range of twelve billion US dollars (BU 2002: 243, 166). When secondary hazards, indirect economic losses, and social disruption are reflected in the damage assessment, broader dimensions of a scenario earthquake can be anticipated for the largest city in Turkey which is the center of the Turkish economy.

3.2 The Development of the Istanbul Metropolitan Area

Istanbul has suffered from major earthquakes several times in its history. There are historical records to mosques and churches collapsing, giant waves forming and ships colliding at the sea, thousands dying and living outdoors from fear or from destruction of earthquakes and following fires.[18] The last of these major earthquakes occurred on July 1894. Even though records indicate only 130 casualties in Istanbul, and 500 casualties around Izmit, speculations about potential earthquakes had reached heights after this earthquake, causing Istanbul residents to sleep outdoors for a period of one month (Alkan 1999: 36).

[17] HAZUS (Hazard US) is a loss estimation tool, developed by FEMA and the National Institute for Building Sciences (NIBS). This GIS-based loss estimation software has the capability of using both deterministic and probabilistic information. It uses four classes of information in order to calculate a probable maximum loss. These are: (1) Map-based analysis (e.g. potential ground shaking intensity); (2) Quantitative estimate of losses (e.g. direct recovery costs, casualties); (3) Functional losses (e.g. reconstruction of critical facilities); and (4) Extent of earthquake induced secondary hazards (e.g. distribution of fires, floods) (Cutter 2001: 29).

[18] See *İstanbul'da 1894 depremi* (Ürekli 1999); *İstanbul depremleri* (Genç and Mazak 2000); and *The Seismicity of Turkey and Adjacent Areas* (Ambraseys and Finkel 1995), for the historic seismicity of Istanbul and its adjoining areas.

Fig. 3.5 Sheltering residents after the 1894 earthquake in Istanbul (Newspaper (Illustration) *Courtesy* of the national information service for earthquake engineering, EERC, University of California, Berkeley)

Despite these records, the city's development shows signs of disregard at the potency of earthquake hazards in the city. This section explores urban planning and development practices that have taken place in the Istanbul metropolitan area, in an attempt to identify their role in the making of the current exposure and vulnerability in the city (Fig. 3.5).

3.2.1 The Development of Istanbul from the Establishment to the Republic

The development of Istanbul from a Greek colonial settlement in BC seventh century into the capital of the Eastern Roman Empire, Constantinople, in the fourth century, took place largely within the city walls of its Historical Peninsula.[19] The city's spatial structure mostly remained unchanged until the Ottoman

[19] The part of the city surrounded with Golden Horn on the north, Bosphorus on the east, Marmara Sea on the south, and Theodosius city walls on the west, is called the Historical Peninsula.

Fig. 3.6 A cartographic image of Istanbul in the 16th century [*Source* Porcacchi (1576)]

Conquest in 1453. Thereafter, Istanbul gained a new identity with development movements and the sheltering of Turks, Greeks, Armenians, and Jews settling from various parts of Anatolia and elsewhere (Kuban 1993: 16). In the sixteenth century, while the Ottoman Empire was in its enlargement period, Istanbul reached the peak of its monumental development. However, in the following centuries, large fires due to the use of timber in vernacular architecture, and growing cultural tensions in the Empire started affecting the spatial structure of the city (Fig. 3.6).

A new socio-economic and spatial period started for Istanbul in the late eighteenth century, The Ottoman Empire joined the world economy with trade agreements and capitulations. The ensuing westernization movement changed the spatial organization of the city. First, the palace and the ruling population abandoned the Historical Peninsula and moved into new Baroque style palaces and mansions on the European side. With this movement, the class structure started to manifest itself in space. A dual business center was formed, as the traditional bazaars remained in the Peninsula and a new non-Muslim mercantile class developed on the European side in Beyoğlu. As new industrial activities brought a big immigration flow to Istanbul, the city's population increased from 250,000 in 1829 to 900,000 in 1914. Single men occupied buildings formed poverty areas transforming Eyüp, Kasımpaşa, and Üsküdar districts into slum settlements (Tekeli 1994: 47). Meanwhile, with the introduction of tram and ferries, Istanbul's growth started to take place in the suburban developments along the banks of the Bosphorus.

Westernization movements had an effect on the urban administrative system, and initiated planning studies in the city. In 1839, Helmuth Von Moltke was commissioned to plan the walled city to provide a continuous street network. With new regulations, construction methods were detailed, and in burnt areas, residential fabric was reorganized in a grid pattern (Çelik 1993: 51–53). The first western style municipality was established in 1854 in Beyoğlu/Pera, with the initiatives of the wealthy Levantines. Within the next decade, a municipal organization was established for the entire city fulfilling modernization by enlarging streets, opening park areas, and providing masonry construction.

Istanbul's urban reform took a new turn with the revolutionary Young Turks from 1908 until the collapse of the Empire. Under the rule of the military and the political organization of this movement, an extensive modernization and city-building program took place, providing public works from sanitary infrastructure to public transportation services, in an attempt to revive the Empire and its capital city in their last years of rule.

3.2.2 1923–1950: Planning in the Nation-State

After centuries of being a capital city to both the Byzantium and the Ottoman Empires, Istanbul lost its important identity as an imperial power with the establishment of the Turkish Republic and the transfer of administrative functions to Ankara in 1923. Proclaimed as a "city of the past"[20] by the official publication of the Republic, *La Turquie Kemaliste* (1943),[21] Istanbul entered into a new socio-economic and spatial transformation period. In less than a decade, its population decreased almost in half (from 1,203,000 in 1919 to 690,857 in 1927). As Tekeli (1994: 51) noted, the city's population had also become more homogeneous: "In 1855, the percentage of Muslims was 44 %, that of non-muslims was 41 %, and foreigners 15 %. In the census of 1927, these percentages were 64, 27 and 9 % respectively."

The change in the social structure and the ongoing fires were quickly shaping the spatial structure in the city. In the traditional imperial seat of the Historical Peninsula, abandoned areas had lost their function and status; they were either left unoccupied, or "had passed into the hands of tradesmen and wealthy families from Anatolia" (Tekeli 1994: 64). The new prestige areas were the European side districts of Beyoğlu, Teşvikiye, Şişli, and Nişantaşı, where the non-Muslim population and the "nouveau riche" lived in the newly constructed brick and stone apartment buildings (Tekeli 1994: 65).

Urban administration and legislation were also modified with the new Republic. In 1930, the *Law of Municipalities* (Belediyeler Kanunu) and the *Law of General Sanitation* (Umumi Hıfzısıhha Kanunu) required all large towns to prepare urban plans. That same year, a new administration system was set up for Istanbul. The city was identified as a single municipality; it was divided into ten sections, and its

---

[20] Translation by Sibel Bozdoğan.

[21] Ankara-Istanbul (1943).

administration was merged with the centrally appointed provincial government. As Alada (1994: 134) argued, the major problem for Istanbul in the early republic era was its' disconnect with the central government due to its historical identity. It can be argued that the joint administration system was an initial step in strengthening the ties of the old and the new capitals, while at the same time controlling the city's planning and development according to the new national ideals. Three years after this new administration scheme, Istanbul entered into a major planning phase, with the introduction of the *Municipality Construction and Roads Law* (Belediye Yapı ve Yollar Kanunu) that required municipalities to prepare their urban plans within five years.

Elgötz's Proposal. Following the new Municipal Law, four[22] European *urbanists* were invited for a competition to prepare plans for Istanbul. Alfred Agache, Hermann Elgötz, and Jack Lambert stayed in Istanbul for a one-month period to prepare their reports. The jury, formed by the municipality, examined their reports and accepted the proposals of the German planner Elgötz, who proposed the development of the transportation circuit and the preservation of historic monuments. As the overarching idea for the plan, Elgötz (1934: 5) wrote: "It is necessary to find a principle, to preserve the exclusive beauty of this city to the far future, such that would integrate the old culture appropriately with the contemporary needs and conditions of civilization.... Within the last years, it was tried to modernize this old soul. While this movement is going on, it is necessary to preserve old monuments carefully and in this way the value of the city can be achieved artistically."[23] Elgötz's plan focused on topics of (a) public transportation (sea transportation and harbors, railroads and stations, airports and streets), (b) zoning, (c) public spaces, (d) old buildings and monuments, and (e) building orders, in which Elgötz (1934) suggested that it was necessary to determine the elevation and construction orders to be able to execute this master plan.

Two of Elgötz's planning ideas considered environmental issues. One of those was under the subject of zoning, in which Elgötz proposed to move heavy industry with its existing factories outside the city walls in order to reduce the effects of air pollution. The second one was his consideration the suitability of the land in his proposal to establish a large airport between the Çekmece Lake and the sea (Elgötz 1934: 14).

Even though Elgötz's proposal was accepted and received positive criticisms by the municipality commission, heavy reactions came from national architect–urbanists, who opposed the idea to provide Istanbul's planning with a competition among foreign architects (Duranay et al. 1972: 72). The proposal also received criticisms for technical reasons, in particular on the subject of the locations of the harbors. At the end, Elgötz's proposal was not transformed into a plan, with the reasoning that it did not provide any solution to direct the city's development.

---

[22] Among them, Henri Prost could not attend due to his responsibilities as the main urbanist of Paris (Tapan 1998: 78).

[23] Author's translation.

Wagner's Studies. After rejecting Elgötz's proposals, the city invited another German urban planner, Professor Martin Wagner, to plan Istanbul. In his planning studies from 1935 and 1938, Wagner (1937) focused on the city's future development and proposed transportation schemes that would connect land with the water of Istanbul's hinterlands. Wagner explained that due to Istanbul's dense population (2,900 people per km^2), the city ought to acquire its needs from its hinterlands, and that the government and the municipality should take necessary steps to develop these areas. Wagner also stressed the potential of migration to Istanbul and its hinterlands, which could then increase unemployment in the city. He argued that Istanbul's hinterland plan should consider ways to prevent migration, such as developing industry outside the city borders (Duranay et al. 1972: 12). At the end, Wagner's studies were not considered satisfactory, leading way to Istanbul's most ambitious planning works.

Prost's Plans. Following failed planning attempts, in 1936, the chief urbanist of Paris, French architect Henri Prost was invited to Istanbul to take charge of planning activities. Despite economic difficulties rising from the Second World War, Prost's work in Istanbul lasted from 1937 to 1950, dominating the development of the city.

Prost envisioned Istanbul in three separate entities. In 1939, his 1/5,000 master plans for the Historical Peninsula and the Beyoğlu regions were put into implementation. In 1940, he prepared a separate master plan for the Anatolian side, which proposed a circulation system along the southern coast allowing the year-round use of suburban neighborhoods between Kadıköy and Pendik. In his reports, Prost (1938: 19–20) detailed necessary actions to implement his plan, including as the foremost principle "aerial photos for topographic maps" and a series of plans ranging from 1/10,000 transportation schemes to 1/2,500 detailed plans for districts envisioned to be developed.

The program of Prost's plans was based on etudes on: (a) railroad and sea transportation; (b) formation of the bazaar; (c) small industry, crafts and their expansion; (d) expansion of industry and commerce; (e) formation of the modern buildings and sanitary orders; (f) development of all neighborhoods forming the Peninsula; (g) archeological researches and principal structures, sites and monuments (Prost 1938: 3).

However, the research did not go on exactly as Prost had anticipated. As Prost's assistant Angel (1987: 36) later wrote: "In general, completing the planning work was hard and complicated. Prost had neither appropriate documents, nor valid topographic drawings. The statistic information was too bad. Information about industrial and commercial activities, demographic and social etudes and documents on land taxes were not complete."[24] Angel (1987: 36) also criticized the implementation period: "A large avenue was proposed to be built at the end of the Atatürk Boulevard…. This coastal road would have continued along the Marmara Sea. It would have formed green areas between the sea and the strolling areas; and

[24] Author's translation.

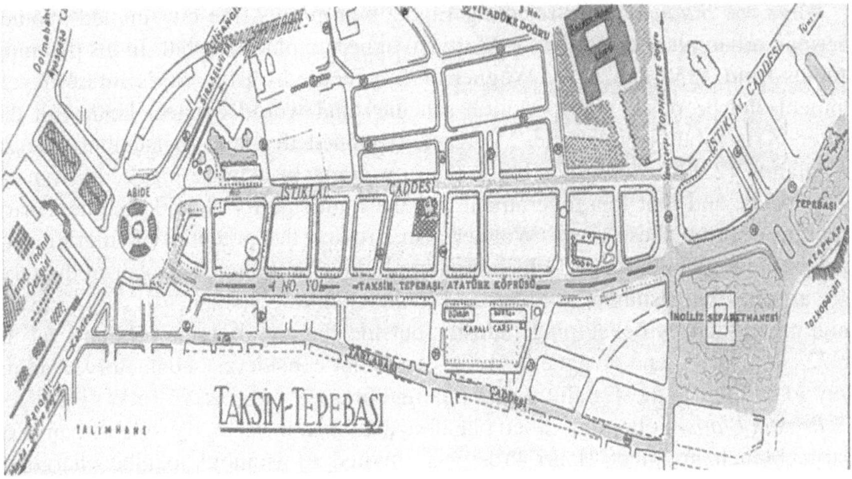

Fig. 3.7 Henri Prost's plan for Taksim, Istanbul, 1937 [*Source* Prost (1938)]

it had to follow the railroad in order to stress the connections between the commercial and residential areas and between beach and the recreation areas. Unfortunately, this road was built near the coast and not only did this prevent the connection of this region with the sea and the use of the coast, but it also created a monotonous view without providing the possibility of any green area along the road."[25]

Despite such difficulties in research and implementation, many of Prost's decisions, including the establishment of major roads and boulevards, the introduction of industry outside the borders of the Historical Peninsula, the reorganization and construction of public squares and stadiums, and the clearing of their surrounding areas were implemented, with the support of Istanbul's administrator, Lütfi Kırdar. Lutfi Kırdar, who was the Mayor and Governor of Istanbul from 1938 to 1949, expressed: "[t]he most important activity achieved during the war period is the Atatürk Boulevard. It was once crowded with narrow streets and old timber houses, but now it is a broad main road passing from the middle of the city, from the Marmara coast to the Golden Horn" (İB 1947: 6–13) (Fig. 3.7).[26]

Prost's planning approach and the clarification of the old and narrow city fabric resembled Baron Haussman's urban renewal works in Paris at the end of the nineteenth century. As Prost had mentioned in his report, he wanted to build large roads and boulevards and destroy the old city fabric, which he considered unsuitable because of dimension and importance. Such a planning approach suited the new Turkish Republic that wanted to "create a new and thoroughly modern nation" (Bozdoğan 2001: 67). As Bozdoğan (2001: 67) argues "[t]hat Le

[25] Author's translation.

[26] Author's translation.

Corbusier lost a major urban commission in Turkey, when he suggested that the Turkish government spare the old wooden houses of Istanbul from any modern intervention illustrates the strength of these republican sentiments." As Corbusier (1949: 230–31) later wrote: "If I had not committed the most strategic mistake of my life in the letter I wrote to Ataturk, I would be planning the beautiful city of Istanbul, instead of my competitor Henri Prost. In this notorious letter, I foolishly recommended to the greatest revolutionary hero of a new nation to leave Istanbul as it was, in the dirt and dust of the centuries."[27]

The Prost planning agenda that was supported by the central government continued until the end of the Lütfi Kırdar administration. However, Prost's Plan affected future planning activity. The implementation of his projects was completed with the construction of the main arteries in the Historic Peninsula from 1957 to 1960; making a large impact on the city's current development.

3.2.3 1950–1960: Planning Faces Gecekondu

As Istanbul's population grew from 700,000 in the 1930s to 975,000 by 1950, Prost's plan started to be widely criticized by the planning and the architectural academic circles, for not foreseeing the city's future and growth potential. A plan *Revision Commission* formed by the Municipality Council and was comprised of representatives of university planning programs, the Union of Architects and Engineers, and the Bank of Provinces[28] examined Prost's plans in detail and concluded with the following:

1. Not only the method used in the preparation of the master plans did not match with the contemporary urban planning approach, but also planning activity had started without any surveys.
2. Due to the lack of existing maps, developed plans had no connection with city topography.
3. As the planning scale changed, there was a detachment with the main idea of the plan.
4. It was observed that the dominant planning approach of the studies was the adornment of the city, instead of finding solutions to residential, transportation, social and economic problems (İB 1954).[29]

The criticisms of the *Turkish Chamber of Architects* (Türk Mimarlar Odası 1960) to Prost's planning approach were similar as they argued: "The planner

---

[27] Translation by Sibel Bozdoğan.

[28] In 1945, the *Law of Bank of Provinces* (İller Bankası Kanunu) was introduced to establish the Bank, which provides local municipalities with technical aid and financial loans to prepare maps, and master and applications plans.

[29] Author's translation.

could not understand the main problems of a very important, complex and beautiful city such as Istanbul. For example, his idea that the city's population would not exceed 800,000 (and why 800,000) shows us that he does not have any information about the natural advantages of Istanbul and about the start of migration to big cities in our country."[30]

Indeed, the housing shortage caused by the loss of old housing stock[31] after Prost development operations, and the population increase by rural to urban migration[32] had initiated different building activities in the city. As Tekeli (1994: 92) noted, as Prost's "reconstruction activities related to the establishment of the urban circulatory system and the beautification of the city was going on, the city was living through a severe housing crises," which resulted in the development of *gecekondu*[33] (squatter) building activity in Istanbul.

Official reactions to migration and illegal housing in Turkey had started as early as 1924 with a "law that authorized municipalities to demolish dwellings built on unowned land" (Danielson and Keleş 1985: 171). In the 1930s, it was proposed that "migration to the cities should be regulated and that those who wish to come to the cities should first obtain a permit" (Heper 1978: 17). Other proposals had included "to levy a special urbanization tax from the new arrivals" and "to create urban activity in the rural areas" (Heper 1978: 17).

The 1930 Municipal Law had authorized municipalities to issue construction permits and the 1933 Municipality Construction and Roads Law had "enabled municipalities to demolish construction which had been started without obtaining a permit" (Heper 1978: 17). But, it was the 1948 *Law Concerning Demolishing of Buildings Without Construction Permits* (Ruhsatsız Yapıların Yıktırılmasına Dair Kanun), which detailed the demolishing process and brought harsh penalties to the squatters, started decades long struggle and court battles involving squatters, law enforcement officers, and politicians.

In Istanbul, gecekondus first emerged closed to industry or manufacturing sites. As Tekeli (1994: 95) described, since the General Sanitation Code had ruled only the very large establishments to locate outside of residential areas, low-income workers were concentrated around manufacturing activity in the city centre north of Beyoğlu, and in the suburbs of the Anatolian side. The first area that was

---

[30] Author's translation.

[31] It has been reported that 1,148 buildings were destroyed during Prost-Kırdar development operations (Tapan 1998: 80).

[32] This internal migration after the Second World War was based on the decrease of rural work demand with agricultural mechanization assisted by the postwar US Marshall Plans. Increase of construction work and establishment of new industry in Istanbul were the main pull factors for the newcomers.

[33] *Gecekondu*, a Turkish word born in the 1940s, means, "Built overnight," and describes the illegally constructed squatter buildings. According to its official description in 1966, gecekondus are "dwellings erected, on the land and lots which do not belong to the builder, without the consent of the owner, and without observing the laws and regulations concerning construction and building" (Karpat 1976: 16).

different from the original small group of shanties and achieved the status of a full gecekondu neighborhood was Zeytinburnu, accommodating 3,200 gecekondus in 1949 (Tekeli 1994: 94).

In the early 1950s, as the Revision Commission that was formed after Prost worked to elaborate Prost's plans, the *Committee of Consultants,* formed mostly by the Revision Commission members followed up their suggestions from 1953 to 1956. The Commission suggested the development of heavy industry, port, and harbor between Pendik and Tuzla along the southern Marmara coastline (one of the highest earthquake risk areas), as well as the expansion of the Golden Horn industry area to the entire Golden Horn. Their suggestions were further developed in the *Plan of Istanbul's Industry Regions* in 1955, giving form to the current industrial pattern in Istanbul. At the same time, the growth of industry along the coast of the Golden Horn accelerated the deterioration of the area, which transformed into a slum settlement for single workers.

As Cansever (1993: 53) wrote, when the native inhabitants left this area, the abandoned buildings were transformed into warehouses, to industrial functions, and to the first shelters of the newcomers to Istanbul, who had not yet built their own squatters. Indeed, this author's interviews with residents in one of these neighborhoods, Fener in 1995 indicated that the area was still regarded as a temporary location for the newcomers, who expressed their wishes to move to other districts in the future (Okyay et al. 1995) (Fig. 3.8).

In 1955, at the end of a five-year period after Prost, Istanbul had reached a population of 1,400,000. Tekeli (1994: 102–103) portrayed that in this period, the

Fig. 3.8 A resident looks at the lens in the slum neighborhood of Fener (Photograph by author, 1995)

spatial structure of the city was characterized by an increase of motor vehicles, uncontrolled urbanization that corresponded with the residential growth and acceleration of the industry, and development outside the municipality borders.

In this period, the increase in population had not only created the gecekondu phenomenon, but also new forms of urban development such as flat ownership and housing co-operatives that led the city's expansion to new areas. In 1948, the newly introduced *Law Encouraging Building Construction* (Bina Yapımını Teşvik Kanunu) facilitated the building of social housing by enabling loans to residents, who were allocated land by the municipality for that purpose (Heper 1978: 18–19). In the following year, the municipality transferred one hundred hectares of public land in the European side to create a new residential neighborhood with four hundred houses (Tekeli 1994: 98). However, "a housing committee meeting in Istanbul calculated that some 5,760 residential units needed to be built in Istanbul annually," and as the market supply could not meet the demand, gecekondus expanded rapidly, creating a dual structure, consisting of legal and illegal housing stock in Istanbul (Tekeli 1994: 98–99).

Menderes' Operations. A new legal and administrative framework had to be provided to deal with the rising problems of urbanization. In 1956, the first *Development Law*[34] (6785 Sayılı İmar Kanunu) and an *Expropriation Law* (6850 Sayılı İstimlâk Yasası), which "provided public institutions with the power to expropriate land for public interest" were approved (Heper 1978: 37). These laws facilitated the start of what is known as the *Menderes Operations*. In the fifth year of his governance, Prime Minister Adnan Menderes became interested in assuming the authority for conducting reconstruction activities in Istanbul, eliminating the significance of the 1958 separation of the provincial and the municipal administrations in the city (Tekeli 1994: 104). The political activity of the government led to the establishment of the *Ministry of Reconstruction and Resettlement* in 1958. That same year, the Bank of Provinces established the *Istanbul Development and Planning Directorate* and in 1959, the *Istanbul Development Regulation* was introduced.

Menderes's desires to make Istanbul the showcase of the developing and the modernized Republic had turned the city into a construction arena. Turgut Cansever reported that during these operations, the road network proposals of Henri Prost's master plan were broadened into larger scale development plans after consultations with the then fashionable transportation engineers, but without making any survey of site conditions (Kuban 1995). Meanwhile, in order to direct planning operations, German planner Professor Hans Högg was invited to the city. Högg helped prepare a transportation-oriented plan, in which he suggested that the

[34] The development law was an important act to facilitate urban development and planning activities, bringing together many laws and regulations under a single legislation. A continuation of the patriarchal relation between the local and the central government, the law required all municipalities to prepare development plans, and to send it for approval to the central government to the newly established Ministry of Reconstruction and Resettlement.

transportation network should relate to the structure of the city, to historical elements, and to city views (Duranay/Heper/Ural 1972: 84).

With the objectives of solving traffic congestion and beautifying the city, the ensuing operations focused on opening large arteries, boulevards, and squares. With the addition of these operations to the prior ones, from 1950 to 1960, about 7300 buildings were expropriated and many historical structures were destroyed (Tapan 1998: 82). As Doğan Hasol (1994: 26) expressed, "In this period, all around Istanbul, a rapid destruction was taking place for the enlargement of roads and the construction of new ones. Vatan and Millet Streets, Sirkeci-Ataköy Coastal Road, the Barbaros Boulevard, Dolmabahçe-Tophane axis, Kemeraltı Street, the main bazaar street in Üsküdar, the upper roads on the Bosphorus, the Beyazıt Square, Karaköy Square and many others had an entire appearance of ruins. While these roads were being opened—whether they had historical or architectural importance—the significance of buildings or their environments were not being considered."[35]

However, it was not only the historical significance of the city that was being disregarded during these operations. These expropriations caused many people to lose their workplaces and to move from their houses, adding to the growth of the city's gecekondus. Meanwhile, with the 1953 *Law for the Encouragement of Construction and Related to Buildings without Licences* (Bina Yapımını Teşvik ve İzinsiz Yapılar Hakkında Yasa) all squatters built up to that date were regularized (Tekeli 1994: 105). Even though harsh punishments to new gecekondu constructions were introduced with this act, according to a survey carried out by the Ministry of Reconstruction and Settlement, in the early 1960s, about 40 % of Istanbul's dwellings qualified as gecekondu areas, and 45 % of the city residents were considered gecekondu inhabitants (Karpat 1976: 11).

3.2.4 1960–1980: Planning the Region

Piccinato's Plan. As the Menderes operations were going on, the newly established *Istanbul Development and Planning Directorate,* under the direction of Cevat Erbel, invited Italian planner *Luigi Piccinato* to direct planning of an Istanbul Metropolitan Area Plan. Called as the *Transition Period Master Plan*, the objective of the two and a half year studies were described as "developing a new organism, which would not only provide new development areas, increase population, fix residential conditions, and manage the city's traffic, but also by settling the city's production tools to more advantageous areas, would link the city's development to economic function" (İB 1962).[36]

---

35 Author's translation.

36 Author's translation.

The 1/10,000 scaled plan was designed for a population of two and a half million people, and it argued that Beyoğlu (European side) and the Historical Peninsula had reached a saturation point and had to be decentralized (İB 1962). The plan discussed the significance of environmental factors at two topics. The first connection to environment was about the development potential between Bakırköy–Florya axis, and its' being exposed to north winds. The second was about the identification of the potential of the Marmara Region, in which "geographical areas (climate, potential of areas)" were mentioned as one of the forces that determined these potentials (İB 1962).

Piccinato recommended that "by a national level policy decision, Istanbul should be converted into a commercial, consumption, cultural and administrative center instead of a manufacturing center and that industry be decentralized" (Tekeli 1994: 126). Ilhan Tekeli argued that Piccinato's plan was different from previous studies as it had attempted to solve Istanbul's problems with a new spatial organization at the regional level (Tekeli 1994: 126). Piccinato's plan had also aimed at transforming the radio-concentric city form into "a decentralized, open and linear system" supported by a "backbone" transportation system (Tekeli 1994: 126).

As this planning period was ending, a military coup in May 1960 ended Prime Minister Menderes' rule. Within the following months, the Ministry of Reconstruction and Resettlement established the *Marmara Regional Planning Office* supported by the Organization for Economic Cooperation and Development (OECD) and the United Nations. Following these developments, despite appraisals by the national and the international planning circles, the Ministry stated that Piccinato's plan had to be revised according to the upcoming regional studies, and it did not approve the plan (Tapan 1998: 84). Following this development, a *Development Planning Directorate* was established in the municipality, dissolving both Högg's office and the departments of the Bank of Provinces, where Piccinato was working.

Transition Period Studies. With the establishment of the Development Planning Directorate, a new planning era started in Istanbul. This *Transition Period* was expected to last until a new plan for the city was introduced. During this period, in alliance with the National Five Year Plan, which proposed to choose superior areas as the starting points for regional development, the Eastern Marmara section of the Marmara Region was selected for development (İB 1966: 23). A pilot project was initiated for the development of the Istanbul-Izmit-Adapazarı axis.

The *Eastern Marmara Regional Plan* projected an approximate population of five million people living in Istanbul by 1980. According to this plan, the most important factors for the preparation of the physical settlement scheme of the Istanbul Metropolis were: (a) its need for future land, (b) development directions, (c) its coast, and, (d) its connection with the region (Duranay/Heper/Ural 1972: 98). According to the plan's physical settlement proposals, there would be a coastal development axis, extending from Büyük Çekmece to Gebze. The plan envisioned green areas along the coastal strips, a regional transportation line and

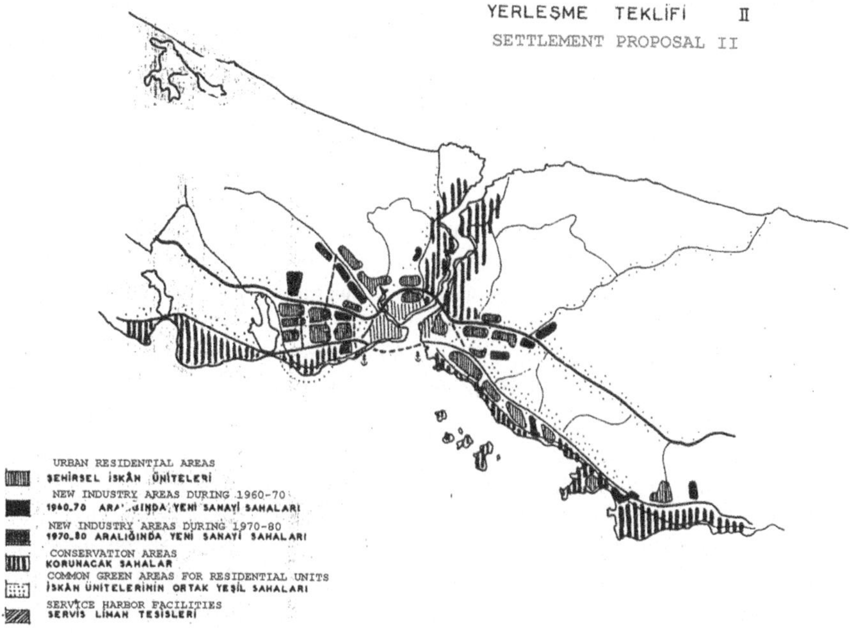

Fig. 3.9 One of the two proposals for the Eastern Marmara regional plan in 1963 *Source* Author's adaptation from: Istanbul Belediyesi (IB 1966)

residences, and that industry would connect with sea only in a number of locations (Duranay/Heper/Ural 1972: 98). The planners had found the regional development axis advantageous because of its "regional structure, protection of the coast, ability to develop in an orderly and dense way, its suitable topography, and link between center and environments" (Duranay/Heper/Ural 1972: 99). However, despite conclusion of this planning study and its adoption by the planning circles, it could not be implemented due to the incompletion of its complementary studies (Fig. 3.9).

Another planning attempt of the Transition Period was the preparation of the *Plan for the Adjacent Areas of Istanbul*, with the goal of providing for the areas in between two municipal boundaries. According to urban theorist, Tekeli (1994: 132), the goal of the plan was to prohibit construction and subdivision of lands in unplanned areas and preventing the development of industry along the Bosphorus and the Kağıthane and Alibey valleys. He argued that it was not easy to implement this plan "because of the existence of the right of referendum" as in one case in a Kağıthane village, where residents "refused to join in with the 'adjacent' area and preferred to subdivide and sell its land within the context of village law and collect urban real estate profits" (Tekeli 1994: 132).

A third planning attempt of the Istanbul Municipality Development and Planning Directorate was the 1964 *Walled City Development Plan*, which proposed to zone the Historical Peninsula. The 1/5,000 scaled plan divided the inner city into six main zones, providing them with different elevations and density (Duranay/Heper/Ural 1972: 100–102). However, the planners did not develop the zoning details of the plan, and following what Tapan (1998: 85) and Tekeli (1994: 138) call a substitute to this plan, a *Floor Regulation Plan for Istanbul* was issued. The Regulation Plan allowed for increases in the number of stories of buildings in the legal housing stock of the city, increasing building height limits from 15.50 to 18.50 m.

The allowance of building height increases helped the newly developing housing supply process of "build and sell" type of construction. Engaged by small entrepreneurs, this process started providing housing supply, particularly for the upper middle-income groups in the already settled sections of the city; either by demolishing old houses or by constructing on gardens and vineyards (Tekeli 1994: 161). This type of process was dominant along the shoreline on the west of the city, between Bakırköy and west of Yeşilköy, as well as on the eastern side between Kadıköy and Bostancı.

Simultaneous to these developments, gecekondus continued to expand from small concentrations distributed on unsuitable land in the city to large scale neighborhoods located either on urban peripheries or outside of the municipal boundaries (Tekeli 1994: 151). According to the *Eastern Marmara Regional Plan* studies, gecekondus occupied 9 % of the land within the municipality borders (İB 1966: 21). However, it was the settlements outside of the municipality borders that were rapidly spreading. From 1950 to 1960, the number of population living in Istanbul's suburbs had multiplied almost by four, and by 1970, by more than fourteen times (Danielson and Keleş 1985: 64).

In 1966, the *Squatter Housing Law* (Gecekondu Kanunu) brought a new perspective to the phenomenon of gecekondus, acknowledging them as both a social and a physical problem. The Ministry developed a fund to provide loans for residents, in order to build and repair houses or to buy land. A second fund was provided to the municipalities to buy and build houses and to provide public services (Heper 1978: 24). However, the law also made clear that rural invasion towards the municipal and the public land was not acceptable, and that squatter settlements would immediately be demolished. The enactment of this law did not solve the gecekondu problem as Turkey's passage into a multiparty political system brought a new dimension to the problem. Michael Danielson and Ruşen Keleş (Danielson and Keleş 1985, 177–179) report that the limited capital and technical resources of the municipalities, unpopularity of the introduced system with squatter residents, and political agendas resulted in amnesties and in the establishment of peripheral municipalities outside the city borders.

The Great Istanbul Master Plan. In 1966, with financial support from the Bank of Provinces, the Ministry of Reconstruction and Resettlement established the *Great Istanbul Master Plan Office.* This office received the plan making authority of all municipalities within the metropolitan area, and with Luigi Piccinato as a consultant, it initiated introductory studies for the Istanbul Metropolitan Area Master Plan between 1966 and 1968 (Güzelsu 1985).

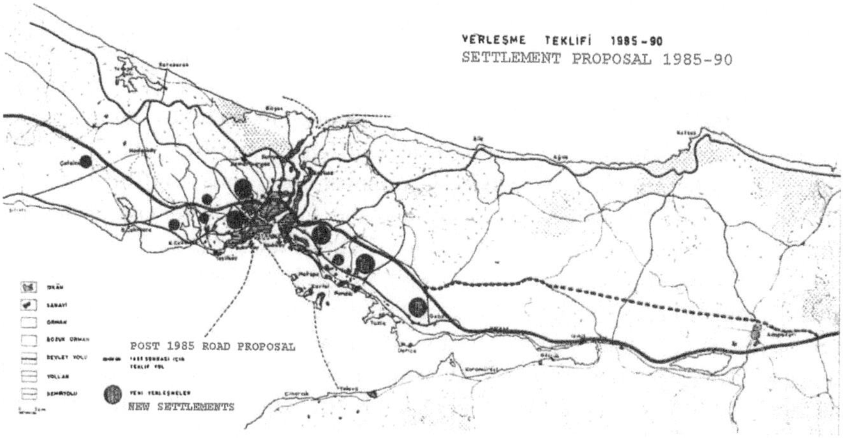

Fig. 3.10 The great istanbul master plan settlement proposal 1971 [*Source* Author's adaptation from: "Büyük İstanbul nazım plan bürosu 1971–1972 çalışmaları" (1971–1972 greater Istanbul master planning studies), in *Mimarlık* 7 (1972): 110–112]

Piccinato believed that if the existing development pattern of the Istanbul Metropolitan Area continued, it would resemble a "poor octopus shaped metropolis with an empty hinterland" (Tapan 1998: 85). In order to change the radio-concentric city form into a linear one, the new planning scheme had to be supported with a long-term regional infrastructure plan, in which the region would no longer be limited with the Eastern Marmara Region, but would include the whole Marmara Region (Tapan 1998: 85). Piccinato proposed that this integration would be possible with speedways and harbors providing rapid sea transport to link the region.

Simultaneous to this proposal, there was a major public discussion on the proposal of building a Bosphorus Bridge. The Chamber of Architects believed that a bridge would change the macro-form of the city, and it would only help residents with private cars, and therefore it would not help solve the transportation problem for the lower and middle classes (Tapan 1998: 85). However, supported by Prime Minister Demirel, the Bosphorus Bridge was proposed in the *Local Belt-Ways Plan* of the Istanbul Municipality, and it was quickly approved by the Ministry, before waiting the completion of studies of the Master Plan (Tapan 1998: 85–86). The construction of the bridge started in 1970 and it was completed in 1973 (Fig. 3.10).

Meanwhile, the Master Plan Office planned new surveys and studies in order to complete the Istanbul master plan. However, the possibility of being financed by the World Bank rushed these studies, and before the completion of the research, in 1971, the 1/25,000 scaled Istanbul Metropolitan Area Master Plan was completed (Tapan 1998: 86; Tekeli 1994: 197). The plan estimated that the population of the Istanbul Metropolitan Area would reach 5,500,000 people by 1990, and that the ratio of population growth on the east side would be higher than on the west side (Duranay/Heper/Ural 1972: 106). The plan adopted an open and linear city form supported by the Bosphorus Bridge and its belt-ways. In this way, the establishment of a second linear axis would facilitate a balanced decentralization decreasing the weight on the coastal axis, as the development moved to the east and to other areas of the Metropolis (Duranay/Heper/Ural 1972: 106). The plan suggested the development of secondary administrative centers and suggested the unification of the Istanbul Municipality and the smaller municipalities under a single metropolitan administration (Duranay/Heper/Ural 1972: 106).

Following the 1971 military intervention, a scientific committee evaluated the Great Istanbul Master Plan and concluded that the plan was not satisfactory due to lack of research, and advised its revision. Refusing the approval of the plan, the Ministry, in cooperation with the Bank of Provinces, set up a Development Planning Bureau for the Istanbul Peripheral Municipalities (Tapan 1998: 86; Tekeli 1994: 200–202).

Istanbul Urban Development Project. As the revision works for the Great Istanbul Master Plan were going on, several articles of the Development Law were modified in 1972. The changes gave authority to the Ministry of Reconstruction and Resettlement to make and to commission plans without the approval of the concerned municipalities (Tekeli 1994: 205). Two weeks after these changes, the Turkish Government and the World Bank signed an agreement for an *Istanbul Urban Development Project,* which would be undertaken by foreign consultants, and to finance and to support the development of five short-term projects [37]of the Great Istanbul Master Planning Office (Tekeli 1994: 205; Tapan 1998: 86–87).

The involved planners proposed to develop the Istanbul Urban Development Project in two phases. Accordingly, in the first phase, settlement models would be developed to decide land use decisions for 1995 (Güzelsu 1985: 83). In addition, priority areas for urban development and gecekondu prevention (for the second phase) would be selected (Güzelsu 1985: 83). Two models developed in the first phase emphasized the city's development either on the east or on the west side of Istanbul. The model emphasizing development on the west side of the city was selected for implementation (Güzelsu 1985: 84). This model aimed at decreasing the pressure on the Historical Peninsula, and suggested to develop a new business center and large housing complexes between the Büyük and the Küçük Çekmece Lakes (Tapan 1998: 87; Tekeli 1994: 207).

However, geological studies conducted after the 1975 plan found out that the plan's main proposal suggesting development between the two lakes was unsuitable (Tapan 1998: 87; Tekeli 1994: 207). The departure of the planning director after political changes, the 1975 census, and the result of the geological studies led

[37] These projects were: (a) improvement of gecekondu areas, (b) development of an urban center on the eastern part of the city, (c) location change of the wholesale market, (d) review of the DAMOC sewage system, and (e) a traffic engineering and control project (Tekeli 1994: 205–206; Tapan 1998: 87).

to a revision of the plan in 1976. The 1976 plan made only slight changes to the plan, increasing the estimated population and suggesting that 70 % of the population would live on the west side and the remaining 30 % would live on the east side of the city (Tapan 1998: 87; Tekeli 1994: 209). Despite geological studies, proposals to promote development between the two lakes were re-introduced in this revised plan. As the director of the planning bureau, Güzelsu (1985: 86–87) later explained: "The city form that was preferred with the 1976 report, follows a policy that supports development on the West Side of the Metropolis rather than on the East Side, and takes into account the need for development between Mahmutbey and the Çekmece Lakes.... In terms of transportation, the selected land use strategy envisions the development of three main east–west corridors and the construction of connecting roads that will link the proposed development between Eminönü, Beyoğlu and the Çekmece Lakes."[38]

This plan received several criticisms from the planning circles. A report of the Istanbul Technical University Institute of Urbanization (İTÜ 1978) found several problems with the plan, ranging from conceptual definitions, accomplished researches, and plan proposals. The university planners found out opposing arguments on the plan report. For instance, while in a paragraph on land uses, the plan report mentioned that "a large portion of the area between the Çekmece Lakes under military control was creating important obstacles for development,"[39] in another paragraph regarding urban form, this obstacle was not mentioned (İTÜ 1978: 8). In another paragraph, investment proposals were introduced for this area as if the previously mentioned problem was already resolved. Again, in another opposing proposal about the Çekmece Lakes, the plan proposed a dense settlement as the desired city form, while in another paragraph it criticized "development in water basins" (İTÜ 1978: 9).

Despite these criticisms, being bound by new studies ordered by the Prime Minister for the development of a second bridge and having "no control or direct participation in the activities of the foreign experts" (Tekeli 1994: 210), the Master Planning Office developed the 1/50,000 scaled *Istanbul Metropolitan Area Master Plan*, which was approved in 1980.

Istanbul Metropolitan Area Master Plan. The Istanbul Metropolitan Area Master Plan was prepared to accommodate a population of 7,100,000 people by 1995. According to the plan, 33 % of the population would reside on the east side of the city, and 67 % on the west (Tekeli 1994: 210).

The goal of the plan was declared as: "To increase the significance of the Istanbul Metropolitan Area for the nation and the world, without losing its identity, and to create functions and services that are necessary for the growth and the development of the metropolis, in alliance with the development of the nation" (Mortan 2000: 71).[40] The plan aimed at: (a) utilizing Istanbul as a world city,

[38] Author's translation.

[39] Author's translation.

[40] Author's translation.

Fig. 3.11 The 1980 approved Istanbul metropolitan area master plan [*Source* Büyük İstanbul Nazım Plan Bürosu (Istanbul)]

(b) preparing mega projects suitable for the mentioned goal, (c) providing an interregional balance, (d) preventing migration, and (e) increasing urban attraction (Mortan 2000: 71).

The plan proposed a hierarchical structure for the central business districts, with the core remaining south of the Bosphorus Bridge, with beltways on the Beyoğlu side, and business areas in the Historical Peninsula (Tekeli 1994: 211). According to the proposed structure, two first class centers would be established within the residential areas of Bakırköy and Kadıköy (Tekeli 1994: 211). Even though, as a general principle for protection areas, the plan had mentioned geologically undesirable areas, secondary centers would be located between Büyükçekmece and Küçükçekmece (on the west side) and in Gebze and in Kartal (on the east side) (Mortan 2000: 77; Tekeli 1994: 211). As for residential areas, Avcılar was one of the areas proposed to be developed as a middle density settlement. In addition, "dense residential areas developing between the two lakes were to be encouraged as mass housing areas," and also in Gebze on the eastern side (Tekeli 1994: 211) (Fig. 3.11).

3.2.5 1980–1999: Planning the Globalizing City

The year 1980 represents an important turning point for Istanbul and Turkey. Following the 1980 military intervention, and a new Constitution in 1982, Prime Minister Turgut Özal introduced a neo-liberal economic policy in Turkey, facilitating economic integration with the world. Accordingly, the state changed its previously highway-focused infrastructure policies into those focusing on telecommunication investments that were necessary for an information society. At the same time, new institutions were established to decentralize administrative structures.

As part of these developments, in 1984 with the *Law of Greater Metropolitan Municipality* (3030 Sayılı Büyükşehir Belediyelerinin Yönetimi Hakkında KHK), a two-stepped administration model was introduced for metropolitan cities in Turkey. According to this new model, each of the districts in the metropolitan area was to have separate elected mayors, along with a greater metropolitan municipality with executive powers. In addition to an increase of municipality resources, with a new *Development Law* (3194 Sayılı İmar Kanunu), development rights were transferred to municipalities. The responsibility of preparation, approval, and application of 1/5,000 scaled master plans and the plans with upper scales were assigned to the Greater Metropolitan Municipality, whereas the 1/1,000 application development plans and smaller scaled plans were to be prepared by local municipalities with the approval of the greater municipality.[41]

This administrative restructuring came at a time, when there was a strong interest on the establishment of a special administrative system for Istanbul and its regions. First, in 1975, the municipalities in the Marmara region, with a focus on common environmental problems, united to form the *Union of Marmara and Bosphorus Municipalities* (Tekeli 1994: 171). Second, in 1979, thirty-two municipalities in the Istanbul metropolitan area formed the *Union of Istanbul Municipalities*. And third, with two decrees of the military regime in 1981, all peripheral municipalities and villages were abolished and merged with the Istanbul Municipality, increasing its population from 2,853,000 to 4,351,000 (Tekeli 1994: 172). And in 1984, with the new administrative model, the *Greater Istanbul Metropolitan Municipality* was established to serve the Istanbul metropolitan area along with the fourteen existing district municipalities.

Despite its advantages, due to the dissolution of past planning offices, the new restructuring of the Istanbul Metropolitan Municipality affected planning operations adversely in the initial stages. Because of the closing of the Master Planning Bureau, the approved 1980 master plan could not be put into application. Meanwhile, because of the inadequacy of the municipal control of urban development and planning activities, the city continued to sprawl to new areas facilitated with the new Bosphorus Bridge. In addition, illegal developments and gecekondus were supported by a series of building amnesties. First, in 1976, all previously built gecekondus were regularized. According to Tekeli (1994: 214), "in a study conducted in 1982 to determine the number of gecekondus built after the building amnesty of 1976, some 208,000 gecekondus or unauthorized buildings were recorded within the metropolitan municipal borders," averaging to 35,000 dwelling per year. Despite this information, in 1983 a general building amnesty

[41] For more on the local administration structure in Turkey, see *Yerinden yönetim ve siyaset* [Local administration and politics] (Keleş 1992).

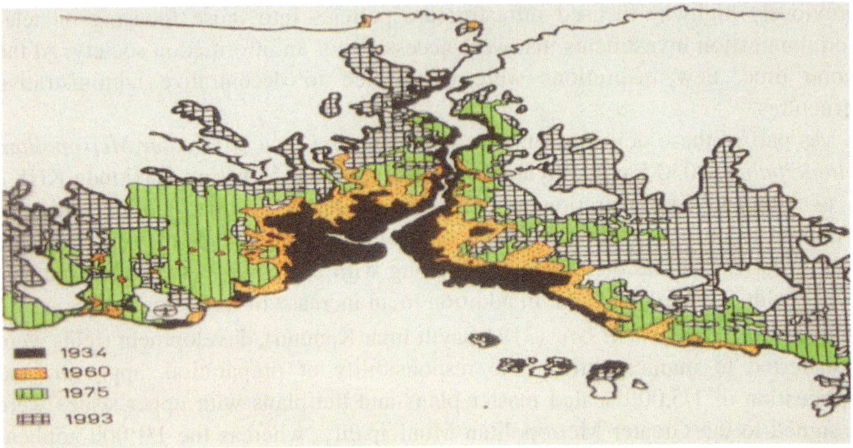

Fig. 3.12 Istanbul's Development 1934-1993 [*Source* Author's adaptation from Kuban (1994)]

was introduced for all illegal buildings in the city with the exception of the Bosphorus area (Fig. 3.12).

In the 1980s, there was a change in the form and the style of gecekondus as the continuous building amnesties gave way to a rising real estate market in these areas. Gecekondus were no more constructed merely as a shelter for the newcomer, but as a tool to bring rental and sales profits. A strong land mafia emerged with this speculative market, and new peripheral municipalities rose in the outskirts of Istanbul.

According to a study of the Provincial Directorate of Istanbul, in 1992, in the Province of Istanbul, 850,000 buildings had permits, 750,000 buildings were previously regularized by building amnesties, and 400,000 buildings were illegal (Sönmez 1996: 140; Mortan 2000: 49). The State Planning Organization had estimated that 40 % of the gecekondus were between 75 and 99 m^2, and 35 % of them were between 50 to 74 m^2 in size (Sönmez 1996: 140; Mortan 2000: 49). According to the same study, 17 % of the gecekondu dwellers had personally established their dwellings in Treasury Land, and 56 % of the dwellers had purchased them from sellers, who had previously gotten public land, indicating the rising real estate market in these areas (Sönmez 1996: 141; Mortan 2000: 49) (Fig. 3.13).

However, it was not only the gecekondu developments that had started to evolve in the 1980s. As a new professional and economic class emerged with the liberalized economy and a globalized social life, new spatial developments surfaced in the city. Reflecting global influences, these spaces varied from high-rise towers to gated communities, and luxury villa developments. Many of the new residential developments were located in areas not open to development, such as in green areas overlooking the Bosphorus, or in forest areas on the northern part of the city. The new upper class and their exclusive developments, located side by

Fig. 3.13 Unplastered half-finished buildings with satellite antennas on the roofs are often encountered in the outskirts of Istanbul, Tuzla, Istanbul (Photograph by author, 2004)

side by gecekondus, brought a new dimension to the social and spatial inequality in Istanbul (Fig. 3.14).

Indeed, in the 1990s, the rapidly growing migrant population had reached such levels that only 37 % of the population in Istanbul was born in the city (Sönmez 1996: 125; Mortan 2000: 51).[42] The new coming populations were either unemployed, or working in temporary or low-skilled jobs. By then, Istanbul had a significant role in the nation's economy, accommodating 40 % of total industry, as well as 75 % of the real estate and financial institutions, bringing 42 % of national tax revenues (Ünsal et al. 2001: 5). However, this generated income has been dispersed in a way that has widened the income gap between Istanbul residents. According to a survey of the State Institute of Statistics in 1994, in Istanbul, the top 20 % of the population with the highest income levels received 64 % of the total generated income in the city, whereas the bottom 20 % received only 4 % of it (Mortan 2000: 46).[43]

Istanbul Metropolitan Area Sub Region Master Plan. In the 1980s, the development of the city had expanded so much that, in 1989 a study of the 1980 master plan revealed that the plan and the existing situation no longer matched (Tapan

[42] According to the 2011 census, the city receives an annual migration of 450 thousand people annually (Turkiye Istatik Kurumu [State Institute of Statistics]; at: http://tuikapp.tuik.gov.tr/Bolgesel/ (August 2012).

[43] According to the latest available numbers in the Turkish State Institute of Statistics in the publication of this book; as of 2001, GDP per capita in Istanbul is $3,063. According to 2010 numbers, 18.7 % of the population in Istanbul has a 60 % risk of poverty. (Turkiye Istatik Kurumu [State Institute of Statistics]; at: http://tuikapp.tuik.gov.tr/Bolgesel/ (August 2012).

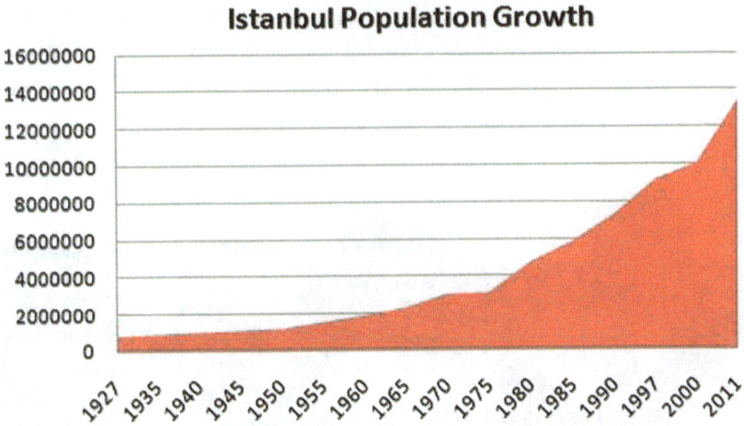

Fig. 3.14 Istanbul Population Growth 1927–2011 (by author) [*Source* Istanbul Buyuksehir Belediyesi (IBB) Istanbul Arastirmalari Merkezi, 1997: Cumhuriyet Donemi Istanbul Statistikleri: Nufus ve Demografi: 1927–1990 [Istanbul statistics during republic period: population and demographics: 1927–1990], Istanbul; Turkiye Istatik Kurumu [state institute of statistics]; at: http://tuikapp.tuik.gov.tr/ (August 2012)]

1998: 88). Following this situation, a new Master Planning Office was established under the Istanbul Metropolitan Municipality. The new planning office developed the 1/50,000 scaled *Istanbul Metropolitan Area Sub Region Master Plan* based on the principles of the 1980 plan. The Istanbul Metropolitan Municipality approved the new plan in 1994 and reapproved it in 1996 following revisions after a political change (Fig. 3.15).

The Metropolitan Istanbul Sub-region master plan estimated a population of 9,000,000 people in Istanbul for 2010, and aimed "to establish a balance between conservation and development for Istanbul as a city that joins with the economic structure of the world and the region" (TRGIM 1995). The Plan had three main strategies:

1. *Rule of Specialization* Within the encompass of Metropolitan Area Sub-region planning the housing-work relations of especially those who are new comers by resolving it in a rational manner and improvement of this relations which were ill defined in the previous structure, within the framework of a plan.
2. *Rule of Ranked Centers* In order to achieve the decentralization of population in the entire Metropolitan Area Sub-Region suggesting wing-attraction centers and ensuing the development of these as primary centers. Achieving the growth of the urban macroform in a linear and multi-centered fashion with a degree of ranking.
3. *Rule of Ranked Density* In accordance with the analysis carried out for the whole of Istanbul, decreasing the sustainable population densities gradually from the centers to outwards and decreasing the mean values (TRGIM 1995) (Fig. 3.16).

Fig. 3.15 New residential villas on the hills of Istanbul on public land that was previously not open to public, Bosphorus, Istanbul (Photograph by author, 2005)

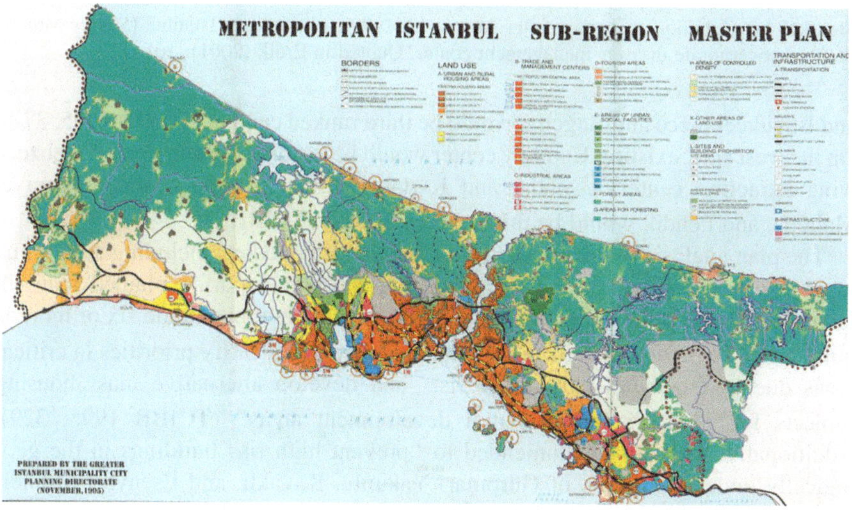

Fig. 3.16 Istanbul metropolitan area sub-region master plan by 1/50,000 Scale (*Source* Greater Istanbul municipality, planning and zoning control general department, city planning directorate, 1995)

According to these strategies, in order to develop the city in a linear fashion and to reduce pressure on existing centers, sub-centers would be developed. Accordingly, on the eastern section of the city, the existing Bakırköy center would be the primary sub-center, Ortaköy-Kavaklı, wing attraction centers, Avcılar, a secondary center,

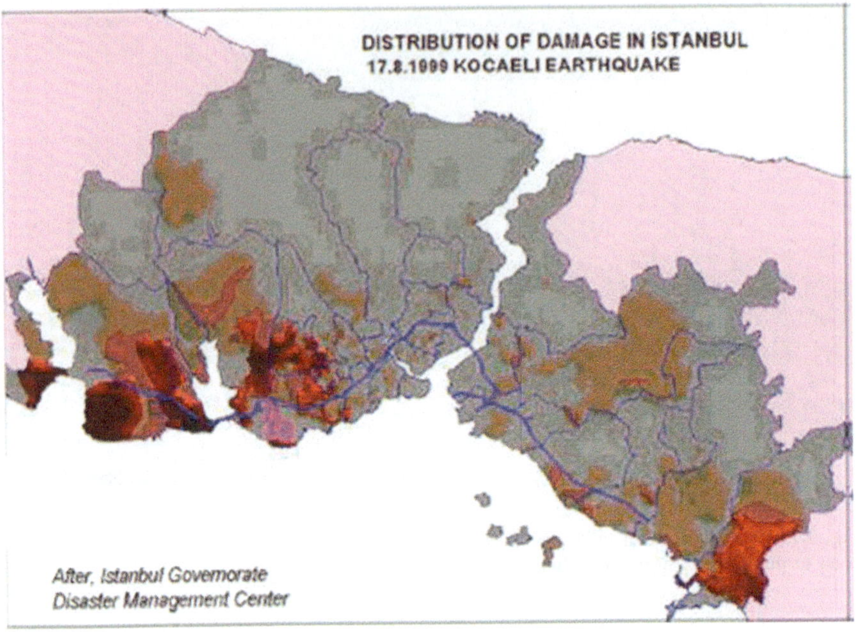

Fig. 3.17 Distribution of damage after August 17, 1999 earthquake in Istanbul [*Source* Map by Istanbul governorate disaster management center, Quoted in Erdik (2001): 16]

and Büyükçekmece and Bağcılar would be third ranked centers (TCİBB 1995: 377). On the west side, existing Kadıköy center would be the primary sub-center, Gebze, a wing attraction center, Üsküdar and Kartal, secondary centers, and Ümraniye, Maltepe, and Pendik would be third ranked centers (TCİBB 1995: 377).

The plan analysis involved a series of geophysical studies, including a first time review of past earthquakes in Istanbul, and the plan report explained that "in Istanbul, the possibility of a damaging earthquake with a magnitude six or more is quite high" (TCİBB 1995: 52). The plan proposed to "identify priorities in critical areas due to Istanbul's earthquake risk, and develop alternative mass housing projects for squatters in residential development areas" (TCİBB 1995: 329). Additionally, the plan recommended to "prevent high rise buildings in the geologically undesirable area of Gürpınar, Yakuplu, Kavaklı, and Esenyurt Municipalities by providing low densities," and proposed "active green areas to provide function so as to protect geologic and archeological sites north of the Küçükçekmece Lake" (TCİBB 1995: 329).[44] In addition to these first-time analysis and proposals reflecting geophysical conditions of the city, the Metropolitan Municipality established the *Geotechnical and Earthquake Investigation Directorate* in January 1997 (Özerol 2001: 116).

[44] Author's translation.

However, these studies were already too late for the informally urbanized city, and settlements extending to geologically unstable areas. When the August 17, 1999 Marmara Earthquake occurred, the problems of the city revealed themselves with destruction and loss of life. According to the Istanbul Governorate records, 981 people lost their lives, 41,180 residences and workplaces were damaged, and 18,162 families needed temporary sheltering (TCİV 2002). With 17,863 damaged buildings (BU 2002: 40), the most affected district was Avcılar, located between the two lakes, where previous planning activities had proposed mass housing and business centers (Fig. 3.17).

3.3 Conclusion

After examining the hazard and risk profile of Istanbul, this study illustrated urban planning and development policies that have taken place in the city, since the establishment of the modern Turkish Republic in 1923, in order to understand their role in the making of the 1999 earthquake disaster and creating the current socio-economic and spatial vulnerability.

This exploration revealed the negative impact of unsustainable urban development on creating the vulnerability of populations. Analysis of the urban development and planning of the Istanbul Metropolitan Area has shown the importance of *problem recognition* as the essential component of creating sustainable and disaster-resistant communities.

It was observed that none of the planning studies (except for the 1996 master planning study) considered Istanbul's past disasters. Even though geological studies that were undertaken during the preparation of the 1976 plan indicated unsuitable settlements around Büyükçekmece and Küçükçekmece Lakes, this problem was overlooked, making its way into future planning studies. Consequently, a commercial center and a large-scale residential development were proposed between the lakes, an area, which had the highest level of destruction in the 1999 earthquake and is again expected to have the highest exposure in a potential future earthquake.

This is only one of the planning decisions that led to the current vulnerability of the Istanbul metropolitan area and its residents. One-dimensional and short term planning activities are the second set of problems Istanbul has faced. Prost's and Menderes's planning activities were one-dimensional in the way that they focused solely on what was referred to as "beautification" efforts, as a way of creating the new Istanbul suitable for the modern Republic. These activities lacked any necessary background studies for a comprehensive planning activity, and did not envision a future growth for the city.

These short-term planning actions, in addition to the lack of national development policies, have resulted in the enormous difficulty of creating sustainable living conditions in Istanbul and other major cities in Turkey. The problems surrounding gecekondus and other irregular developments with no construction parameters and in disobedience with zoning ordinances are much larger than being solely physical issues. Not only were long-term sustainable political, social, and physical solutions were ignored to the needs of the incoming populations, but also short term political considerations have extended this massive problem to such levels that gecekondus that had initially started as modest self-provided shelters have turned into a major real estate market sustaining an informal economy. While the first coming populations had improved their socio-economic conditions, only the informal real estate groups have profited since the 1980s with the immense flow of populations. Contrary to earlier immigrants, most of the low-skilled newcomers have found themselves unemployed, as the economy was transforming from industrial to service based sectors, creating very wide socio-economic gaps between Istanbul residents and causing social unrests. In addition to being socially vulnerable, these populations are also vulnerable to the effects of natural disasters, given that they live in poorly constructed structures and do not have the means to upgrade, move, or insure in order to protect themselves; stressing the need for a new urban and risk management strategy in the city.

References

Alada A (1994) Cumhuriyet döneminde İstanbul'un yönetimi [Istanbul's administration during republic period], in: Istanbul, vol 8. Tarih Vakfı, Istanbul, pp 133–138

Alkan MO (1999) Sansür, jurnal ve söylentiler arasinda '1894 İstanbul depremi' [The 1894 Istanbul earthquake amid censor, news, and rumors]. Cogito vol 20. Yapı Kredi Yayınları, Istanbul pp 33–41

Ambraseys NN, Finkel CF (1995) The seismicity of turkey and adjacent areas: a historical review. Eren, Istanbul, pp 1500–1800

Angel A (1987) Henri Prost ve İstanbul'un ilk nazım planı [Henri Prost and Istanbul's first master plan]. Mimarlık 1 (Istanbul)

Ankara-Istanbul. La Turquie Kemaliste 47(1943):38–39. Quoted in Bozdoğan S (2001) Modernism and nation building: Turkish architectural culture in the early republic, vol 67. University of Washington Press, Seattle and London

Bibbee A, Gonenc R, Jacobs S, Konvitz J,Price R (2000) Economic effects of the 1999 Turkish earthquakes: an interim report. Working paper 247, organisation for economic co-operation and development, economics department, ECO/WKP (2000)20. http://www.oecd.org/turkey/1885266.pdf

Boĝaziçi University (BU) (2002) Earthquake risk assessment for istanbul metropolitan area. Final report. Kandilli Observatory, Department of Earthquake Engineering; Boĝaziçi University, Earthquake Research Institute, Istanbul

Boĝaziçi University (BU), Istanbul Technical University (ITU), Middle East Technical University (METU), Yıldız Technical University (YTU) (2003a) Earthquake master plan for Istanbul. Metropolitan Municipality of Istanbul, Geotechnical and Earthquake Investigation Department, Planning and Construction Directorate, Istanbul

Bolt B (2004) Earthquakes. 5th edn, W. H. Freeman and Company, New York

Bozdoğan S (2001) Modernism and nation building: Turkish architectural culture in the early republic. University of Washington Press, Seattle

Brauch HG (2003) Natural disasters in the mediterranean (1900–2001): from disaster response to disaster preparedness. In: Hans GB, Liotta PH, Antonio M, Paul R, Mohammed S (eds) Security and environment in the mediterranean. Conceptualizing security and environmental conflicts. Springer, Berlin, pp 863–906

Bruneau M (1999) Structural damage. In: Preliminary reports from the Kocaeli (Izmit) Earthquake of august 17, 1999. Reconnaissance report. University of Buffalo, Multidisciplinary Center for Earthquake Engineering Research, New York. http://mceer.buffalo.edu/research/Reconnaissance/turkey8-17-99/ScawPrelim.pdf

Cansever T (1993) Ülke ölçeğinde İstanbul'u planlamak" [Planning Istanbul in national scale]. Istanbul 4 (Istanbul: Tarih Vakfı)

Çelik Z (1993) The remaking of Istanbul: portrait of an Ottoman city in the nineteenth century. University of California Press, Berkeley

Columbia University (CU) (2002) International urban Planning studio: disaster resistant Istanbul, final report (Unpublished). Columbia University, Urban Planning Program, New York

Cutter, Cutter S (ed) (2001) American hazardscapes: the regionalization of hazards and disasters. Joseph Henry Press, Washington

Danielson M, Keleş R (1985) The politics of rapid urbanization: government and growth in modern Turkey. Holmes and Meier, New York

Duranay N, Gürsel E, Ural S (1972) Cumhuriyetten bu yana İstanbul planlaması. [Istanbul's planning since the republic]. Mimarlık 7:65–109 (Istanbul)

Elgötz H (1934) İstanbul şehrinin umumi planı [General plan of the city of Istanbul].Ahmet Sait Matbaası, Istanbul

Erdik M, Durukal E, Biro Y, Siyahi B, Akman H (2001) Earthquake risk to buildings in Istanbul and a proposal for its mitigation. Departmental report no: 2001/16. Boĝaziçi University, Department of Earthquake Engineering, Istanbul

Erdik M, Siyahi B, Şeşetyan K, Demircioĝlu M, Akman H (2003) Current Situation. In: BU [Boĝaziçi University]; ITU [Istanbul Technical Univer-sity]; METU [Middle East Technical University]; YTU [Yıldız Technical Univer-sity]: Earthquake master plan for Istanbul (Istanbul: Metropolitan Mu-ni-cipality of Istanbul, geotechnical and earthquake investigation department, planning and construction Directorate), pp 14–107

EQE International (1999) Izmit, Turkey earthquake of August 17, 1999 (M7.4). An EQE briefing. http://www.eqe.com

Genç M, Mazak M (2000) İstanbul depremleri: fotoĝraf ve belgelerde 1894 depremi [Istanbul earthquakes: the 1894 earthquake with photographs and documents]. İGDAŞ, Istanbul

Güzelsu K (1985) Nazım plan büroları çalışma ve deneyimleri" [Studies and experiences of master plan offices] In: Çubuk, M (ed) Türkiye'de metropoliten alan planlama deneyim ve sorunları kolokyumu [Colloquim on experience and problems of metropolitan area planning in Turkey]. Mimar Sinan Üniversitesi, Sosyal Bilimler Enstitüsü, Istanbul 75–104

Hasol D (1994) İmar mı? yıkım mı?" [Development? or demolishment?] Yapı 151:25–27 (Yapı-Endustri Merkezi, Istanbul)

Heper M (1978) Gecekondu policy in Turkey: an evaluation with a case study of Rumelihisarüstü squatter area. In İstanbul With the assistance of Butler MH, NT Butler. Boĝaziçi University Publications Boĝaziçi University Press, Istanbul

İstanbul Belediyesi (İB) (1947) Yenileşen İstanbul: 1939–1947 neler yapıldı [Revitalizing Istanbul: what has been done 1939-1947], İstanbul Belediye Matbaası, Istanbul

İstanbul Belediyesi (İB) (1954) Revizyon komisyonu raporu [The revision commission report]. İstanbul Belediye Matbaası, Istanbul. Quoted in: Duranay N, Gürsel E, Ural S (1972) Cumhuriyetten bu yana İstanbul planlaması. [Istanbul's planning since the republic]. Mimarlık 7:81–82 (Istanbul)

İstanbul Belediyesi (İB) (1962) İstanbul nazım planı için ana hatlar [Main features for Istanbul master plan]. İstanbul Belediye Matbaası, Istanbul. Quoted in Duranay N, Gürsel E, Ural S

(1972) Cumhuriyetten bu yana İstanbul planlaması. [Istanbul's planning since the republic]. Mimarlık 7:87–93 (Istanbul)
İstanbul Belediyesi (İB) (1966) Pilot bölge ve İstanbul bölgesi yerleşme planı [Pilot area and Istanbul area settlement plan]. İB Plan ve İmar Planlama Genel Müdürlüğü, Bölge Planlama Dairesi, Istanbul
JICA (Japan International Cooperation Agency), IMM (Istanbul Metropolitan Municipality), PCI (Pacific Consultants International), OYO (OYO Corporation) (2002) The study on a disaster prevention/mitigation basic plan in Istanbul including seismic Microzonation in the republic of Turkey. Draft final report summary (JICA)
Karpat K (1976) The Gecekondu: rural migration and urbanization. Cambridge University Press, Cambridge
Keleş R (1992) Yerinden yönetim ve siyaset [Local administration and politics]. Cem Yayınevi, Istanbul
Kuban (1993) Koloni şehrinden imparatorluk başkentine. [From a colonial town to capital of an empire]. İstanbul, vol 4. Tarih Vakfı, Istanbul
Kuban D (1994) Şehrin gelişimi [Development of the city]. İstanbul Ansiklopedisi vol 4. Tarih Vakfı, Istanbul, p 545
Kuban D (1995) 1950'lerde İstanbul'da Menderes imarı [Menderes development in Istanbul in the 1950's], in İstanbul Ansiklopedisi. Tarih Vakfı, Istanbul. Quoted in Tapan, Mete, 1998: "İstanbul'un kentsel planlamasının tarihsel gelişimi ve planlama eylemleri" [Planning actions and the historical development of Istanbul's urban planning], in: 75 yılda değişen kent ve mimarlık [The changing city and architecture in 75 years]: 80–83. Türkiye İş Bankası and Tarih Vakfı, Istanbul
Le C (1949) Le Corbusier ile mülakat" [Interview with Le Corbusier by S. Demiren] Arkitekt 19:11–12, 230–231. Quoted in Bozdoğan S (2001) Modernism and Nation Building: Turkish Architectural Culture in the Early Republic, 67. University of Washington Press, Seattle
Mortan K (ed) (2000) İstanbul: bir sosyo ekonomik değerlendirme [Istanbul: a socio economic evaluation]. T.C. İstanbul Valiliği, Istanbul
Okyay I, Gencer E, Henden B, Özdemir D (1995) Farklı ve uygulanabilir bir kentsel koruma ve geliştirme modeli: Fener örneği [An applicable urban preservation and redevelopment model: case of Fener], Paper presented at the Üçüncü Kentsel Koruma ve Yenileme Kolokyumu, Mimar Sinan Universitesi, Istanbul
Özerol U (2001) Proceeding in İstanbul ve deprem [Istanbul and earthquake].İstanbul Büyükşehir Belediyesi (İBB), İstanbul Kültür ve Sanat Ürünleri, Istanbul, pp 111–120
Parsons T (2004) Recalculated probability of M≥7 earthquakes beneath the sea of Marmara, Turkey. J Geophysl Res 109, B05304. doi:10.1029/2003JB002667
Parsons T, Toda S, Stein R, Barka A, Dieterich J (2000) Heightened odds of large earthquakes near Istanbul: an interaction-based probability calculation. Science 288(5466):661–665
Porcacchi T (1576) Constantipoli [Constantinople] In: Porcacchi, Tommaso: L'isole più famose del mondo descritte da arretino e intagliate da con l'aggiunte [The world's most famous islands described by Thomaso Porcacchi of Castiglione with the addition of engravings by Girolamo Porro of Padova] (Venice: Simon Galignani and Girolamo Porro) (With permission from Biblioteca Nazionale Marciana: Venezia)
Prost H (1938) İstanbul'un nazım planını izah eden rapor [The report that explains Istanbul's master plan]. İstanbul Belediyesi Matbaası, Istanbul
Smith K (2001) Environmental hazards: assessing risk and reducing disaster, 3rd edn. Routledge, New York
Sönmez M (1996) İstanbul'un iki yüzü: 1980'den 2000'e değişim [Two faces of Istanbul: transformation from 1980 to 2000]. Arkadaş, Ankara
Tapan M (1998) İstanbul'un kentsel planlamasının tarihsel gelişimi ve planlama eylemleri [Planning actions and the historical development of Istanbul's urban planning]. 75 yılda

değişen kent ve mimarlık [The changing city and architecture in 75 years]. Türkiye İş Bankası and Tarih Vakfı, Istanbul pp 75–88

Tekeli İ (1994) The development of the Istanbul metropolitan area: urban administration and planning. IULA-EMME and YTU, Istanbul

Türk Mimarlar Odası (1960) Yurdumuzda imar çabaları [Development attempts in our country], report, Ayyıldız Matbaası, Ankara, Quoted in Duranay N, Gürsel E, Ural S (1972) Cumhuriyetten bu yana İstanbul planlaması [Istanbul's planning since the republic], 93–95: Mimarlık 7:65–109 (Istanbul)

Turkish Republic Greater Istanbul Municipality (TRGIM) (1995) Istanbul metropolitan area sub-region master plan by 1/50.000 Scale, Brochure. TRGIM, Planning and Zoning Control and Construction General Department, City Planning Directorate, Istanbul

Turkiye Cumhuriyeti İstanbul Büyükşehir Belediyesi (TCİBB) (1995) 1/50.000 ölçekli İstanbul metropoliten alan alt bölge nazım plan raporu [1/50,000 scaled Istanbul metropolitan area sub-region master plan report]. TCİBB, Planlama ve İmar Daire Başkanlığı, Şehir Planlama Müdürlüğü, Istanbul

Turkiye Cumhuriyeti İstanbul Valiliği (TCİV) (2002) İstanbul afet yönetim merkezi [Istanbul disaster management center]. Briefing

United States Geological Survey (USGS) (2000) Implications for earthquake risk reduction in the united states from the Kocaeli, Turkey, earthquake of August 17, 1999. U.S. geological survey circular 1193. U.S. Department of the Interior and U.S. Geological Survey. Denver

Ünsal F, Erbaş E, Cavuşoğlu E (2001) Social cohesion and spatial segregation in globalisation era: the case of Istanbul. Paper presented at the sixth international metropolis conference, Rotterdam

Ürekli F (1999) İstanbul'da 1894 depremi [The 1894 earthquake in Istanbul]. İletisim Yayınları, Istanbul

Wagner M (1937) İstanbul ve havalisinin plânı" [Plan of Istanbul and its vicinity]. Arkitekt 1–12 (Istanbul)

References

Chapter 4
Post-Disaster Planning and Risk Management in Istanbul

> Even in such a sensitive municipality, we are also a unit under the department of the environment. The institutions do not have any authority, and they are not up-to-date about this matter.
> *M. Bilgin.*
> *(Disaster Management Project, Kadikoy Municipality).*
> [Author's translation].

Following the two Marmara earthquakes and the heightened possibility of a major earthquake in the Istanbul region, there has been a widespread awareness, in Turkey, of the significance of disaster preparedness. The 1999 Marmara earthquakes were followed by a series of disaster risk management activities by international organizations, the central and local governments, academic circles, and the public itself.

As previously mentioned in Sect. 3.1.2 of this book, immediately after the 1999 Marmara earthquakes two significant studies, *A Disaster Mitigation Basic Plan Including Seismic Microzonation* by JICA–IMM, and *Earthquake Risk Assessment of the Istanbul Metropolitan Area* by BU assessed the potential earthquake risk in Istanbul. Following these studies and the establishment of a *National Earthquake Council* and its *National Earthquake Strategy Report* in 2002, the Istanbul Metropolitan Municipality requested the preparation of the *Istanbul Earthquake Master Plan* (IEMP) by a consortium of four universities. This document laid the basis for identifying activities and responsible authorities, as well as preparing an action plan for disaster risk mitigation and management in Istanbul.

This section examines planning and risk management after the 1999 Marmara Earthquakes. It starts by examining risk and urban management in Istanbul and Turkey following the 1999 earthquakes, including bringing in results of the author's empirical research in three district municipalities within the Metropolitan Municipality of Istanbul, which investigated disparities between risk management practices in districts with different social and economic backgrounds. This chapter, secondly, presents institutional, legal, and planning proposals of the Istanbul Earthquake Master Plan and Urban Transformation Projects undertaken by the Istanbul Metropolitan Municipality. The chapter ends by providing a summary of the results of these studies and bringing out lessons learned for a successful risk management and a more sustainable urban development in the metropolitan city of Istanbul.

E. A. Gencer, *The Interplay Between Urban Development, Vulnerability, and Risk Management*, Mediterranean Studies 7,
DOI: 10.1007/978-3-642-29470-9_4, © The Author(s) 2013

4.1 Risk and Urban Management Following the 1999 Earthquakes

At the time of the 1999 earthquakes, the disaster management structure in Turkey had an abundance of central governmental agencies with a complicated arrangement of authorities. In addition, most of these organizations, as well as the laws and regulations related to disaster management, were focused on post-disaster actions rather than pre-disaster mitigation and preparedness.

The dominating laws on disasters were the 1958 *Law of Civil Defence* (7126 Sayılı Sivil Savunma Kanunu), and the 1959 *Law on Precautions and Aid Regarding all Types of Disasters that Impacts the Community* (7269 Sayılı Umumi Hayata Müessir Afetler Dolayısıyla Alınacak Tedbirlerle Yapılacak Yardımlara Dair Kanun). According to these laws, when a disaster is declared by the Ministry of Public Works and Settlements, authority is transferred to provincial and district governors, providing them with the "sole authority with powers of commanding all public and private and even military resources and means, property, all vehicles and man-power included" (Balamir 2001: 210). However, these laws contain only a few articles related to disaster preparedness and mitigation, such as those related to "dissemination of disaster related information to the general public," or "mandatory relocation of a whole neighbourhood because of eminent danger with the Council of Minister's decisions" (Kabasakal et al. 2003: 220). The 1959 law and its by-law give authority to prepare and approve disaster area plans in district, provincial, and metropolitan municipalities to the Ministry of Public Works and Settlement, where a *General Directorate of Disaster Works* exists (Saĝlam et al. 2003: 420).

Immediately following the 1999 earthquakes, new revisions and decrees were added to the laws, and new agencies were established to support disaster management in Turkey. One of them is the *Directorate of Civil Defence* (Sivil Savunma Genel Müdürlüğü), under the Ministry of Internal Affairs and the Turkish Emergency Management Directorate (Türkiye Afet Yönetimi–TAY), responsible for organizing, training, and management of civil defence in disaster prone areas. According to the *Law of Civil Defense* (7126 Sayılı Sivil Savunma Kanunu), in the provincial level responsible bodies are the centrally appointed provincial and district governors, whereas municipalities have no administrative role, rather they are only responsible of implementing the requests of the public administrative bodies (Kabasakal et al. 2003: 224).

In Istanbul, two disaster management agencies were established after the August 1999 earthquake. These are the *Governorship Disaster Management Centre* (Afet Yönetim Merkezi–AYM), and the *Metropolitan Municipality Disaster Coordination Centre* (Afet Koordinasyon Merkezi–AKOM). The Governorship Disaster Management Centre (AYM), renamed in 2011 with a new directorate as *Province Disaster and Emergency Directorate* (Il Afet ve Acil Durum Mudurlugu–AFAD), coordinates disaster management activities in the Province of Istanbul, and it also has district offices. Despite the fact that the

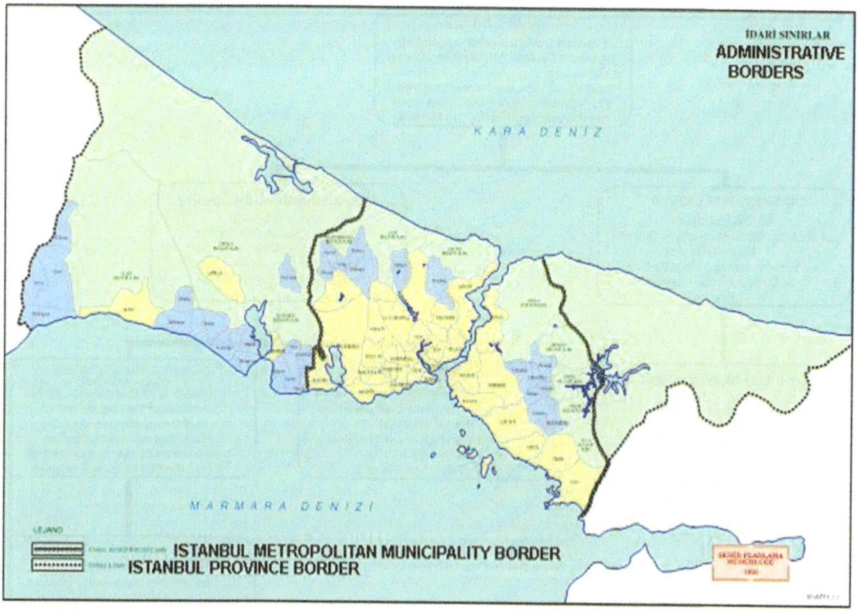

Fig. 4.1 Istanbul administration borders. *Source* Author's adaptation from İstanbul Büyükşehir Belediyesi (IBB), 1995: *1/50.000 ölçekli İstanbul metropoliten alan alt bölge nazım plan raporu* [1/50,000 scaled Istanbul metropolitan area sub-region master plan report]

provincial government has a higher authority in disaster management, it is financially more limited than the Metropolitan Municipality; therefore, it is in need of assistance by the central government, military, and other provinces.

The Metropolitan Municipality Disaster Coordination Centre (AKOM) provides internal coordination between the organizations within the Metropolitan Municipality, and it does not have district offices. Both the Istanbul Metropolitan Municipality and the AFAD report to the Governorship on a monthly basis. This complicated disaster management structure of Istanbul is a result of its complicated administrative structure. Figure 4.1 explains the planning and disaster management authorities of the Province of Istanbul, the Metropolitan Municipality, and the District Municipalities as of 1999, corresponding to Istanbul's administration chart in Fig. 4.2.

According to this complex administrative structure, at the time of the 1999 earthquakes, there were twenty-seven district municipalities, seventeen provincial municipalities, and one independent district municipality within the borders of the Istanbul Metropolitan Municipality. This situation created several problems both for the metropolitan municipality and the district municipalities located along the metropolitan city border, inhibiting the preparation of comprehensive plans and giving full power to independent municipality districts. In the case of one independent municipality, Sultanbeyli, there was full power on the hands of land mafia, who transformed the district from a village to a squatter development and decided

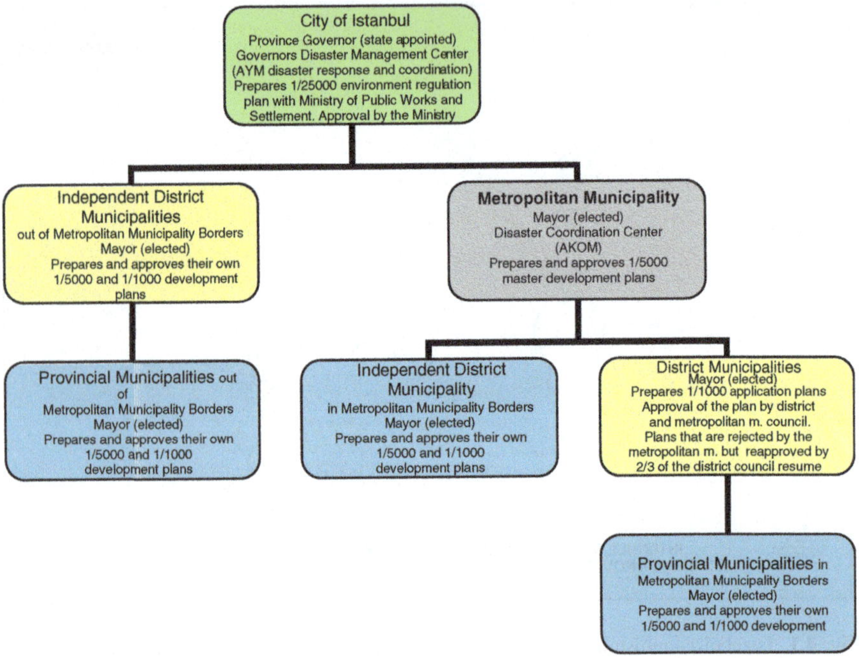

Fig. 4.2 Istanbul administration chart (by author). *Source* Ebru Gencer, 2004: "Sustainable planning for disaster mitigation in Istanbul," paper presented at Prevention consortium web conference, World Bank Wisline Web Meetings

on its independence with elections. This situation gave power to squatter developments to form their own municipalities without having to obey any laws and plans of the Metropolitan Municipality.

The author's empirical research, during 2003 and 2004, in three district municipalities in Istanbul in order to investigate how the complex urban administration system in the city influenced disaster risk management, and to observe whether there were disparities in the way the expected earthquake risk was managed in local districts, exposed differences in levels in actions and strategies of the three selected municipalities in research.[1] According to this research, a striking difference seemed to arise from the organization of municipalities in these districts, and particularly from their budgetary allocations. In Turkey, the largest part of municipal budgets consisted of income from taxes. Municipalities with low income communities lacked adequate income for acquiring experienced and

---

[1] Three district municipalities—Kadıköy, Zeytinburnu, and Avcılar—within the borders of the Istanbul Metropolitan Municipality were selected for the purpose of this investigation, based on their proximity to the earthquake zone and differences in their socio-economic structures. In the selected municipalities, interviews were made to address an open-ended questionnaire comprised of twenty-six questions related to each municipality's and district residents' actions on disaster risk management. For further information on this research, please see Gencer (2011).

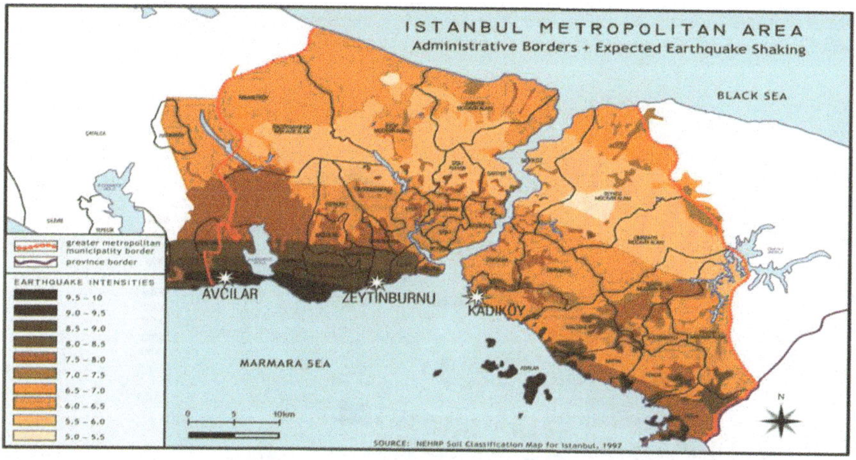

Fig. 4.3 Selected districts in research. *Source* Author's adaptation of *Expected Earthquake Shaking Map of the Istanbul Metropolitan Area*, based on 1997 *NEHRP Soil Classification Map for Istanbul* 2002, in: *International Urban Planning Studio: Disaster Resistant Istanbul*. Final report (New York: Columbia University, Graduate School of Architecture, Planning and Preservation, Urban Planning Program)

specialized staff and to engage in risk management; and the already disadvantaged communities were further neglected due to the lack of financial and technical resources of their local administrations (Fig. 4.3).

For instance, in Kadikoy[2] municipality, the interviews indicated a clear awareness of the potential earthquake hazard, and significance given to the earthquake despite the fact that the building stock in Kadıköy is estimated to be one of the least vulnerable in Istanbul. On the other hand, interviews in the Zeytinburnu[3] Municipality revealed a different picture than that of Kadıköy. Despite the fact that most of the buildings' reinforced concrete exhibited corrosion and several building owners self-implemented building strengthening, the municipality had no control over these buildings, as most of them were either squatter developments or did not have building permits. In addition, there was strong opposition to the input of the municipality from those who benefited from illegal construction, an opposition that came in the form of attacks to the municipality personnel. On the other hand, after experiencing the heaviest destruction in Istanbul from the 1999 earthquake, Avcılar's[4] municipality has made significant changes to its urban planning activities. However, it suffered from

[2] The district of Kadıköy has the highest GDP per capita of the three locations. It also has the highest education rates and lowest annual growth rate among the three districts.

[3] Zeytinburnu, the first squatter district of Istanbul has the highest density and lowest education rates of the three districts.

[4] The district of Avcılar has the highest annual growth rate of the three districts despite its poor soil conditions. It also has the lowest economic level of the three.

Fig. 4.4 A land-use plan at Zeytinburnu Municipality (Photograph by author, 2004)

not having an adequate budget for risk management and from unresolved regional issues due to its location along the borderline of the metropolitan municipality, the neighbouring independent municipality of Büyükçekmece, and the provincial municipality of Esenyurt that caused a chaotic situation in planning and earthquake preparedness (Fig. 4.4).

An unprecedented change to the complex urban administration structure came in July 2004, as a special temporary decree was enacted for Istanbul and Izmit, enlarging the borders of the Istanbul Metropolitan Municipality to match that of the Province of Istanbul. With this decree, the four municipalities outside of the metropolitan municipality borders and the one independent district municipality within the borders of the metropolitan municipality were placed under the authority of the Metropolitan Municipality, increasing the number of municipalities under the

Metropolitan Municipality of Istanbul to thirty-two, and the number of provincial municipalities to forty-one.[5]

This change in the administrative structure in Istanbul is one of several changes related to planning, urban administration, and risk mitigation that have been undertaken since 1999. Following the change in the administrative structure of Istanbul, a programme for a local administration reform has been undertaken by the central government, initiated in 2005 and supported by the UNDP and the European Commission. The local administration program supports the advancement of local administrations in Turkey by "(1) strengthening the capacity of both central and local administrations to formulate and implement reform policies and initiatives, (2) improving budgetary procedures and service performance in selected pilot administrations and 3) improving the efficiency and effectiveness of human resources (through training)" (UNDP Turkey 2006). As part of this program, a new legislation on *Local Authorities* (5393 Sayılı Belediye Kanunu) passed in 2005. Simultaneously, the World Bank provided a major loan for the development of the *Istanbul Seismic Risk Mitigation and Emergency Preparedness Project*, with the objective of improving the institutional and the technical capacity of the City of Istanbul for disaster management and emergency response, strengthening critical public facilities for earthquake resistance, and supporting measures for better enforcement of building codes and land use plans.[6] Responsibility for the implementation of this project was given to a special unit established under the Governor's Office in Istanbul.

Meanwhile, as part of seismic assessment and rehabilitation projects, the Metropolitan Municipality of Istanbul followed up on micro-zoning projects of JICA, and prepared a 1/5,000 scaled geological map of the area within the borders of the metropolitan city. Later, the Municipality initiated a pilot building assessment project in which trained civil engineers assessed buildings in Zeytinburnu neighbourhood in Istanbul in order to provide specific measures, such as whether the buildings should be demolished, repaired, or strengthened in order to withstand a potential earthquake. This project was executed after the establishment of new regulations regarding construction practices. These regulations were the *Decree of Supervision of Construction Processes* (Yapı Denetimi Hakkında KHK) in April 2000, and the *Decree of Proficiency in Constructional Professions* (İnşaat Mesleklerinde Ehliyet KHK) in May 2000, both of which were transformed into the *Building Supervision Law* (Yapı Denetimi Kanunu) in July 2001 and to the *Regulation on Structures to be Built in the Disaster Region* (Deprem Bölgelerinde Yapılacak Binalar Hakkında Yönetmelik) in March 2006.

According to these new regulations, building permits are to be signed by architects or civil engineers representing building control firms that have received government licenses. In addition, the Ministry of Reconstruction and Resettlement

---

[5] İstanbul Büyükşehir Belediyesi (2006).

[6] World Bank (2005).

and the Ministry of Industry agreed to standardize building materials, and to 'discipline' those who do not obey the new standards with strong punishments.[7]

In addition to regulations on construction practices, a *Decree on Compulsory Building Insurance* (Zorunlu Deprem Sigortası KHK) was introduced in December 1999.[8] With this decree, the *Turkish Catastrophe Insurance Pool* (TCIP) was set up to implement compulsory earthquake insurance in Turkey. According to the 2004 World Bank reports, since the beginning of the programme, "insurance penetration for catastrophic coverage has more than tripled" in Turkey, making TCIP "the largest government insurance programme in the world, providing coverage to about 2 million Turkish homeowners (about 16 per cent of the eligible housing stock)" (Pusch 2004: 75). According to another report, "in Istanbul the programme achieved insurance penetration on the order of 30 per cent" (Gurenko et al. 2006: xiv). However, what these reports lack is the identification of residents who are able to purchase the obligatory insurance, as the author's study on district municipalities reveal, the number of residents purchasing insurance varies greatly depending on their socio-economic conditions.

A measure of the TCIP programme was to prohibit earthquake insurance eligibility of "buildings constructed after 27.12.1999 without any construction licence," in an attempt to support legal housing purchase in Turkey (TCIP: 4).[9] For a similar reason, the government introduced a mortgage system, adding to the new housing credit systems issued by banks.[10] In both systems, the main idea is to provide earthquake-resistant and legal housing. However, the new *Mortgage Law* (Passed in 2007 as Law No. 5582 Konut Kredisi Yasası), has contradictions with the newly issued construction laws, as it permits the eligibility of mortgage for housing at the project and at construction phase, which conflicts with the requirement of systematic control for building permits. Another consideration of the law is that it does not impose a restriction on the age of buildings eligible for mortgage, even though new earthquake construction ordinances came into place only in 1999.

The new law also provides funds for retrofitting. However, even though both the Municipality's pilot seismic rehabilitation project and other projects undertaken by academic circles have suggested the cost-effectiveness of retrofitting for

---

7 "Yapı sektörüne deprem düzenlemesi" [Earthquake codification to construction sector], in: Yapı (Istanbul), 2006, 294 (May): 17.

8 This legislation was revised in May 2012, making it mandatory for all buildings with permits to have compulsory earthquake insurance. A new article in this law explains that "In the event an earthquake disaster, the State does not have the obligation to offer loans or build new dwellings to those who did not have compulsory earthquake insurance" (Dogal Afet Sigortalari Kurumu (DASK) [Natural Disasters Insurance Institution]; at: http://www.dask.gov.tr/300.html (August 2012).

9 This condition changed with the revised law in 2012. The new law, however, declares that DASK has the right not to provide insurance to dwellings that were built against the laws and projects (ibid.).

10 "Mortgage dosyası" [Mortage file], in: *Arkitera*; at: http://www.arkitera.com, 3 November 2005>.

the most common type of reinforced-concrete apartment structures in Istanbul (Smyth et al. 2004), there has not been a big interest in retrofitting from private home owners. This is partly because until the modifications in 2012, flat ownership laws required all residents to agree to building strengthening.[11] In addition, it is widely argued that retrofitting a single building will not be successful on lots with attached houses, or in lots where the required distance between buildings is insufficient, putting the risk of each building on its neighbour (Uyaroğlu 2005: 27). On the other hand, in a field survey measuring Turkish homeowner's willingness-to-pay for earthquake measures, it was found that "the role of group dynamics, trust and fairness" played a significant role in earthquake mitigation investment of Istanbul residents; and that the existence of a prior retrofitting of a neighbourhood building made a significant effect on neighbours (Öncüler 2002). Regardless of this analysis, instead of undergoing a retrofitting process, many Istanbul residents have taken steps either to relocate to newly built housing complexes in lower earthquake risk locations on the city outskirts, or to make arrangements with construction firms to demolish their existing buildings (which are usually not higher than six stories) and to rebuild taller ones, in exchange for new apartments.

In 2001, another legal coping strategy regarding the predicted earthquake hazard was introduced by the Ministry of Reconstruction and Resettlement with a draft bill on *Development and Urbanization* (İmar ve Şehirleşme Kanun Tasarısı Taslağı). The bill brought a new terminology to the existing planning documents with the introduction of *Disaster Maps* (Afet Haritaları), which were defined as: "Maps that are prepared in different types and scales, and that compose a whole with their reports as one of the data groups that lay basis to planning, and identify all disaster dangers that can occur in planning areas, and include preventions and proposals related to the reduction of disaster losses and risks" (TCBİB 2005: Article 3). In identifying planning principles, the draft bill explained that "in order to reduce disasters, disaster maps and risk administration reports are to be taken as a principle in planning works"[12] (ibid.,: Article 7) (Figs. 4.5, 4.6).

However, even though the draft bill tried to incorporate disaster mitigation into the planning agenda, its language and complex definitions brought with it many criticisms by the *Turkish Chamber of Planners* (TMMOB SPO) and the *Istanbul Earthquake Master Plan* (IEMP) report team. The draft bill was criticized for not having conferred with the newly issued legislations, and for continuing to create a

[11] According to the 2012 revisions to the *Flat Ownership Law* (634 Sayili Kat Mülkiyeti Kanunu), obstacles to retrofitting are removed. Accordingly, if there is a court rule stating that a building needs strengthening, this can be carried on without the need for all residents to agree. Building residents will collect retrofitting assessment as part of monthly maintenance and residents will no longer be able to make repairs or modifications in their own units that may damage the building's main structure, without the 4/5 approval of the building residents. "634 S.lı Kat Mülkiyeti Kanunu—Son Eklenen Şerhler" [Number 634, Flat Ownership Law—Latest Articles], in: Kat Turk Hukuk Sitesi (Turkish Law Site); at: http://www.turkhukuksitesi.com/serhler.php?kid=66 (August 2012).

[12] Author's translation.

Fig. 4.5 A new housing complex in Beykoz, Istanbul (Photograph by author, 2005)

lack of public participation in the planning process (TMMOB SPO 2005), the dependency of preparing development plans solely based on population criteria without taking into consideration urban dynamics (Sağlam et al. 2003: 446–448).

4.2 Istanbul Earthquake Master Plan and Urban Transformation Projects

As new laws and legislations related to urban and risk management were being enacted, the appointed Istanbul Earthquake Master Plan team proposed two plans to integrate risk reduction into physical planning activities.

Earthquake Mitigation Plan for Istanbul. One of the two projects, proposed by the Istanbul Technical University and the Middle East Technical University team, and named as the *Earthquake Mitigation Plan for Istanbul* (EMPI), is envisioned as a framework to coordinate all mitigation measures "to enhance safety and total quality of life in the city" (Gülersoy et al. 2003: 262).

The EMPI has three components. The first component is the *Contingency Plan,* which is described as "the overall plan to coordinate all documents related to risk sectors, to identify risk management measures, the actors, supervision methods, and the protocols to be drawn between responsible bodies, specifying the lines of action"(Gülersoy et al. 2003: 263). The second component of the EMPI is the *Action Plan*, referring to "methods of immediate intervention in rehabilitation

Fig. 4.6 A self-built house on the hills of Pendik, Istanbul (Photograph by author, 2004)

areas to coordinate property owners and inhabitants, to allow public and private partnerships, with special public powers to synchronize resources and physical development" (Gülersoy et al. 2003: 263). And, the third component of the EMPI is the *Support and Research Activities*, some of which are accomplished in earlier stages. Among these activities are promotion campaigns, public relations, raising resources, legal provisions, administrative coordination, and preparation of protocols, data engineering and other research (Gülersoy et al. 2003: 263).

The EMPI recommends the accomplishment of the following activities for a successful implementation of the Master Plan:

1. Information dissemination and promotion campaigns
2. Formation of action platforms with private sectors and NGOs
3. Administrative cooperation and coordination protocols
4. Cooperation of related parties in risk sectors through protocols
5. Tendering of project packages described in the contingency plan
6. Initiation of pilot action plans
7. Formulation of legal and administrative changes required and monitoring
8. Procurement of resources for implementation
9. Public relations and information engineering (Gülersoy et al. 2003: 263–264).

The EMPI brings a non-conventional proposal, focusing on tools rather than the end product of a development plan. However, the strategy of dividing earthquake risk into multiple parts and attempting to deal with each package separately might bring coordination and responsibility chaos in the implementation phase, to the previously disordered institutional structure. The detailed version of this proposal gives a clearer picture of this possibility, as same risk factors are identified in different subjects without identifying which agencies will be responsible, and creating the potential of an excessive use of time and money in implementation.

On the other hand, the plan gives importance to information dissemination and public participation, both of which have been lacking from previous planning exercises. However, it should be noted that at the time of this planning proposal, the chaotic management structure in Istanbul provided autonomy to areas (such as independent districts), where public votes were the ruling factor in urban development. It is yet to be seen how the new restructuring of urban management will limit development liberties, and how a piece-by-piece planning proposal of this type can be applied without the influence of power structures in the city.

Strategic Plan for Disaster Mitigation in Istanbul. For the Istanbul Earthquake Master Plan, the second working group with Boĝaziçi and Yıldız Technical Universities, developed a *strategic plan*[13] with the primary goal of diminishing "the hazardous effects of a possible earthquake in Istanbul," and which is "supported by a secondary goal of improving the quality of the natural and urban environment" (Ökten et al. 2003a: 195). The *Strategic Plan for Disaster Mitigation in Istanbul* (SPDMI) focused on (a) identifying the problems and the potentials of the Istanbul Metropolitan Area, (b) developing a "road map" with strategies, planning instruments, and priorities, and (c) examining current legal and institutional issues, and proposing recommendations (Ökten et al. 2003a: 195–214, 2003b).

For the planning of Istanbul, the team suggested a "three-fold road map" consisting of macro- and mezzo-level strategies, and micro-level implementation. According to their proposals, as *macro-level strategies,* a new planning level, *National Strategic Plan*, should be introduced by the State Planning Organization (SPO) in order "to indicate the spatial basis of social and economic development" in Turkey (Ökten et al. 2003a: 198). As a second step, a mandatory *Regional Plan* will be prepared by SPO's regional offices and should define the identity for Istanbul. And finally, as a third step of macro-level strategy, the *Master Plan* for the Istanbul Metropolitan Area and the *SPDMI* should work together to create "disaster-resistant areas" (Ökten et al. 2003a: 199–200). Based on high-risk areas, the SPDMI proposed a vision for a settlement plan incorporating regional growth. The proposed settlement plan is based, as a first principle, on the decentralization of settlements in forest, water-basin, and geologically and topographically unsuitable areas to provide a healthy development for Istanbul (Ökten et al. 2003b: 340–341). The SPDMI proposed that a rapid-train system, which was previously proposed in the 1995 master plan, should be extended to facilitate this decentralization. As a second principle, the SPDMI proposed the identification of suitable risk reduction strategies in urbanized settlement areas (Ökten et al. 2003b: 340–341).

---

[13] The team stresses that the SPDMI is a strategic plan, which is based on the principles of "(1)defining urban goals on the basis of current dynamics; (2)the permanent dialectic of goals-projects-repercussions; and (3)public and private agents acting in concert at all the stages of preparation and implementation." (Borja and Castells 1997: 160).

In this step, the team identified priority working areas that are based on problematic and potential areas related to legal status of their initial development, urban functions, population densities, and hazard probabilities. According to their findings, highest areas of priority are: (a) historic areas including both the Historical Peninsula and Galata and (b) the earliest upgraded areas whose initial status were illegal, such as Zeytinburnu. Secondary areas of priority are: (a) planned areas on the west side of the city, such as Bakırköy and Yeşilköy, (b) mass housing areas between the Büyük and Küçük Çekmece Lakes and close to Avcılar, and (c) upgraded areas in Küçük Çekmece, west of Avcılar, and the entire periphery of Istanbul. The final priority areas are: (a) planned areas such as those that lie between Kadıköy and Pendik on the eastern Marmara coast and (b) other mass housing areas (Ökten et al. 2003b: 296–334) (Fig. 4.7).

Upon the identification of priority work areas, *mezzo-level strategies* are developed in four stages including: on-site research, development of tentative strategic plans, following joint working sessions with local officials, public groups, and NGOs, and finally with the production of the final strategic plans.

After mezzo-level strategies, the SPDMI proposes three types of micro-level implementation plans. These are:

1. *Urban Redevelopment Ignition Areas* (Kentsel Dönüşümü Ateşleme Alanları) are defined as areas "which bare a strong potential and enthusiasm for redevelopment and for initiating radical changes at metropolitan scale." It was suggested that urban redevelopment (kentsel dönüşüm) in these areas should be provided in a comprehensive planning approach compatible with upper level plans.
2. *Local Redevelopment Areas* (Yerel Dönüşüm Alanları) are suggested to be areas that may be "lacking development dynamics," but where it is believed that the outcome of redevelopment with rational and effective planning strategies can produce good results.

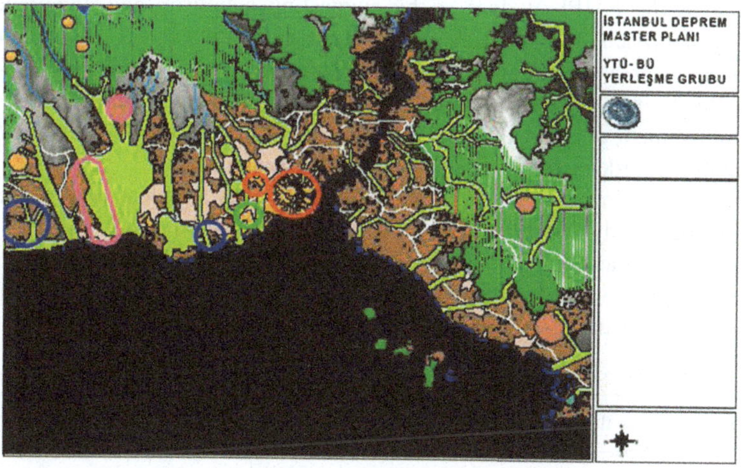

Fig. 4.7 Priority work areas defined by the SPDMI. *Source* Ökten et al. (2003b: 334)

3. *Land Readjustment Areas* (Yeniden Düzenleme Yapılacak Alanlar) are areas with minor problems and lower risk in relation to urban facilities and quality of structures. Community participation is identified as a key element in land readjustment in these areas, in order to make minor changes and increase urban quality (Ökten et al. 2003a: 200–201).
4. In relation to this road-map, the SPDMI team developed specific implementation proposals for problematic areas. The team proposed the use of reinforcement and micro-renovations for risk reduction in planned areas. In upgraded unplanned areas, short-term reinforcement and long-term urban renewal projects were suggested. For unplanned areas that retained illegal status, the SPDMI proposed a topographic reevaluation. According to the SPDMI, in case these areas were found threatening to the sustainability of the city, they would be located to new areas; and in case of the declaration of their suitability, renewal and regeneration strategies would be developed as social projects (Ökten et al. 2003a: 202).

In addition to the physical planning proposals, the SPDMI also proposed institutional and legal changes to the current urban and risk management system. The team advised that regional plans should be mandatory and that two institutional bodies should be empowered in order to provide an "independent control mechanism" (Ökten et al. 2003a: 210). In terms of disaster management, the SPDMI proposed the establishment of a regional scale reorganization, which would include metropolitan municipalities and regional seismic commissions, and that would allow "effective community participation" (Ökten et al. 2003a: 210). In addition, "various applications of 'exchange of property rights' are proposed as alternative compensation mechanisms in planning operations" (Ökten et al. 2003a: 210).

The SPDMI team captures a larger picture in combining the goal of earthquake risk reduction with that of improving the quality of life than its counterpart team proposal of the EMPI. It does so by suggesting a holistic and comprehensive planning approach with multiple level plans compatible with each other, and proposing necessary legal and institutional changes. On the other hand, there is a big challenge in the implementation of their strategy. Their plan largely relies on the establishment of national strategies to create multiple development centers in the country in order to reduce immigration to Istanbul.

Additionally, in urban redevelopment projects, it is not clear how community participation will be established in projects that mandate the movement of populations, or at other times, how local dynamics will be affected with massive redevelopment projects.[14] Local governments are expected to have a major role in establishing such details and providing consensus and financial means for action.

[14] Indeed, some of the initial projects, and most notoriously that of the urban transformation projects in Yedikule—Kumkapi neighborhood in the Historical Peninsula, which required the relocation of a large Roma population have caused much debate and protests from the residents and professional planning circles.

Urban Transformation Projects. With the rule of an ambitious Mayor and the support from the central government, the Istanbul Metropolitan Municipality's actions following the Istanbul Earthquake Master Plan have been far from providing consensus with the public. Indeed, the Mayor's vision of mega-projects has brought a much heated debate in planning and architectural circles, as well as among many Istanbul residents.

Istanbul's Mayor, Kadir Topbaş, initiated his vision for mega-Istanbul projects in 2002. The Mayor's initial ideas symbolized his desire to return the city back to its Islamic Ottoman heritage, with proposals of building two giant statues, one of a whirling dervish, and one of Fatih, the Conqueror Sultan of Istanbul, on the Kadıköy seaport. His following ventures were the initiation of megasize commercial redevelopment projects, Galataport and Haydarpaşa Port, with foreign investments. And finally, with the schemes of the Dubai Towers residential project, of which the slogan is "We are coming to change your life style," the Metropolitan Municipality neglected any prior planning decisions and Istanbul ordinances.

These projects, from proposal phase to bidding, and following lawsuits by professional chambers, were initial signs as to how the Mayor of Istanbul, in cooperation with the central government, wanted to execute urban transformation in Istanbul. Following up on the urban redevelopment project proposals of the Istanbul Earthquake Master Plan team, the Metropolitan Municipality first opened *Urban Transformation Directorate* under its Geotechnical and Earthquake Investigation Department. The initial concept behind the "urban transformation" projects was to follow up on the Zeytinburnu Pilot Seismic Assessment project by building mass housing projects for about three thousand buildings that were assessed as high-risk, and to extend seismic assessment to nine other districts.[15]

Soon after, The Metropolitan Municipality extended the concept of "urban transformation" to large-scale commercial and residential development, and established the *Istanbul Metropolitan Planning and Design Center* (IMP) under BIMTAŞ, a consultancy firm working for the Municipality. In 2006, the IMP announced an international competition for project areas in Küçükçekmece and Kartal, with the objectives of "placing Istanbul within the world's metropolises, and increasing life quality in urban areas".[16,17] Among the invited six international architectural groups, Zaha Hadid's proposal for *Kartal Lower Center and Kartal-Pendik Coastal Zone Urban Transformation Project* and Ken Yeang's proposal for *Küçükçekmece-Avcılar Inner and Outer Beach Area Urban Transformation Project* were selected for implementation. Hadid's project envisages high office buildings, residential blocks and a marina lining the north to south transportation

---

[15] İBB (2007).

[16] Author's translation.

[17] "Dünya mimarların'dan İstanbul için projeler" [Projects for Istanbul from the "world's architects"], in *Yapi* 293 (April 2006):11, Istanbul: Yapı-Endüstri Merkezi.

axis, and social and cultural centers to the east of this axis.[18] Yeang's project's objective is defined as "attaining maximum social, physical and economic environmental benefit".[19] By taking into account the earthquake risk, Yeang proposes detached residential development to the north and the south of the Küçükçekmece center, in addition to a tourism complex along the coastal area.

However, this competition brought with it criticisms from the public who argued that officials had not resorted to their opinions for the projects, and that they were feeling anxious for potential social unrest and anticipated migration in the years to come (Işıngör 2006: 39). In addition to residents of the project areas, professional circles were upset by the Mayor's announcement: "We resorted to the world's most famous architects in areas where Turkish architects do not have enough experience and accumulation…"[20] (Ekinci 2006: 38). As Oktay Ekinci, the former president of the Chamber of Architects, explained in 2005, the Mayor had discussed his wish to open a competition for the Küçükçekmece area with the Chamber of Architects, and upon the notice of the Chamber that the coast-land development ordinances should be applied in such a project, the Mayor implied that he would abandon his idea for the competition (Ekinci 2006: 38). Ekinci does not think that it is only a coincidence that at the time the competition was introduced, the government initiated changes to regulations concerning development in coastal areas (Ekinci 2006: 38).

Meanwhile, in order to legalize and implement these "urban transformation" projects, a *Draft Bill Related to Transformation Areas* (Dönüşüm Alanları Hakkında Kanun Tasarısı) was prepared and passed by the Turkish Parliament in 2006. The Istanbul Metropolitan Municipality (IMM) insisted on the necessity of this bill in order to continue on redevelopment projects on seismic assessment areas (İBB 2007). However, the Union of Turkish Planning Schools (TPOB) and the Chamber of Planners (TMMOB ŞPO) argued that not only has the government not made necessary preparations for the earthquake, but also terminated the previously established National Earthquake Council. In addition to this, professional groups argued, the government took advantage of the "earthquake" concern to pass this bill, which is only a cover up to support capital gain for certain groups (TPOB 2006; TMMOB ŞPO 2006b, 2007). As the TPOB (2006) stated: "The draft bill does not recognize local groups, it does not propose participation models and more importantly it does not mention any arrangement to reform social and economic conditions of the population living in transformation areas. If the draft bill is legalized in its current state perceiving transformation solely as a physical arrangement, arising conditions will provide new possibilities and opportunities to

---

[18] "Altı mimarın İstanbul'a ilişkin çözüm önerileri" [Solutions proposed by six architects for Istanbul]. *Yapi* 294: (May 2006): 63–75, Istanbul: Yapı-Endüstri Merkezi.

[19] "Altı mimarın İstanbul'a ilişkin çözüm önerileri" [Solutions proposed by six architects for Istanbul]. *Yapi* 294: (May 2006): 70, Istanbul: Yapı-Endüstri Merkezi.

[20] Author's translation.

a specific capital sector, while displacing the poor and pushing them to worse conditions."[21]

Simultaneous with the urban transformation competition winners, the IMP also announced the 1/100,000 scaled *Istanbul Province Environment Regulation Plan* (İstanbul İl Çevre Düzeni Planı) in 2006. Upon examining the plan, the TMMOB ŞPO (2006c) applied to the judicatory for its cancellation with the reasons that the plan is contrary to laws, regulations, public good, urbanization principles, and planning techniques. Indeed, the plan summed up the Municipality's urban development activities by declaring its main goal as: "placing Istanbul to its deserving place within the metropolises at the global level, and making it more competitive in the international market," rather than creating a livable urban environment for its residents (TMMOB ŞPO 2006c). The professional Chamber of Architects and Urban Planners criticized the plan for disregarding the Istanbul Earthquake Master Plan report proposals, for being prepared as a land use plan rather than a strategy plan consistent with upper scale planning activities, and for placing significant land uses in the Küçükçekmece Water Basin in addition to other geologically unsuitable areas (TMMOB ŞPO 2006a, c). Despite previous planning decisions to limit development towards the northern parts of the city and to develop a linear light rail mass transportation system, the plan also proposed a new highway on the north side of the existing highway (TMMOB ŞPO 2006a, c) with the Municipality's desire to build a third bridge across the Bosphorus despite protests and criticisms from professional circles. Three years after its proposal, and after several rejections and revisions, the plan was approved by the Metropolitan Municipality Council in 2009.

Accordingly, the new vision for Istanbul is identified as: "developing it in accordance with the environmental, social and economic sustainability principles, while preserving its own cultural and natural identity, and to transform the city into an information society that will have the power for global competition and a high quality of life".[22,23]

4.3 Conclusion

This chapter illustrated post-earthquake urban risk management in Istanbul after the two Marmara earthquakes in August and November of 1999. Many initiatives were introduced following the 1999 earthquakes to prepare for and to minimize the impact of an expected earthquake in Istanbul; such as the preparation of the comprehensive *Istanbul Earthquake Master Plan* (IEMP), the establishment of regulations that increase building construction quality, the imposition of

---

[21] Author's translation.

[22] Author's translation.

[23] İBB (2009).

construction control, and required mandatory earthquake insurance. On the other hand, investigations indicate challenges in the implementation of these risk management strategies in the current socio-economic and administrative state of the polarized city. In addition to self-managed coping strategies, low-income groups have also been at a disadvantage from a complex urban administration system which has not provided a common public policy, and in which disparities have been witnessed in actions of risk reduction.

Among the new urban and risk management initiatives, the Istanbul Earthquake Master Plan is a significant study, because it has brought together multiple data and has laid a strong base for the existing situation. In addition, it has identified the problems that lie on the way for earthquake risk management, and produced physical, socio-economic, and educational proposals in a comprehensive and multidisciplinary effort.

On the other hand, the plan also holds a big challenge for its implementation. First and foremost, it is not clear if the idea of developing "two different yet parallel" strategies is to find ways to integrate them. If so, who will be the responsible body for such an undertaking? Or, is the idea to select appealing components of each strategy for implementation?

Recent actions by the Istanbul Metropolitan Municipality and the central government indicate that such an approach is being undertaken; as parts of the proposals suggesting urban transformation projects have been put into practice, despite the recommendations that such projects should be executed coherently with upper scale plans and with long-term development projects. These actions remind pre-1999 planning decisions and urban management strategies in Istanbul. For this reason, it may be suggested that for the success of risk reduction activities in Istanbul, an independent, apolitical body should be formed, including academicians and scientists, who have prepared the IEMP studies. Moreover, as the *SPDMI* proposed, risk management actions for Istanbul should be supported by national and regional plans providing physical and social development programs for the entire country, in order to reduce internal migration and to accomplish a long-term and sustainable development for Turkey. These plans should be supported with a better financial model for local governments that will improve the current out-dated budgetary system, and help them with technical capacities to implement planning and development activities. The ongoing legal reforms should provide a better administrative and planning system, in which responsibilities are enhanced and clarified decreasing confusion of authority by multiple agencies. They should be supported with a better financial model for local governments that will improve the current outdated budgetary system and help them with technical capacities to implement planning and development activities. They should also provide a stronger ground for public participation of citizenry groups, non-governmental organizations, universities and professional chambers in the planning and development processes. These suggestions and IEMP's proposals can be implemented by the national and local governments, stressing once again that the success of urban planning and risk management actions and policies not only lies on providing sustainable solutions to urban dynamics, but also on their execution with the strong support of policy-makers, who believe in the equitable and just growth of urban areas.

References

Balamir M (2001) Recent changes in Turkish disasters policy: a strategical reorientation. In: Kleindorfer P, Sertel M (eds) Mitigation and financing of seismic risks: turkish and international perspectives. Kluwer Academic Publishers, Dordrecht, pp 207–237

Borja J, Castells M (1997) Local and global: the management of cities in the infor-mation Age. In Belil M, Benner C (eds) United nations centre for human settlements. Earthscan, London

Ekinci O (2006) Kartal ve Küçükçekmece'deki 'iş kazası' [Work accident in Kartal and Küçükçekmece]. Yapı 294:38 Yapı-Endüstri Merkezi, Istanbul

Gencer EA (2011) Risk management strategies for the predicted earthquake hazard in Istanbul. In: Brauch HG, Spring OU, Kameri-Mbote P, Mesjasz C, Grin J, Chourou B, Dunay P, Birkmann J (eds) Coping with global environmental change, disasters and security threats, challenges, vulnerabilities and risks. Springer, Berlin, pp 655–667

Gurenko E Lester R, Mahul O, Gonulal SO (2006) Earthquake insurance in Turkey: history of the Turkish catastrophe insurance pool. World Bank, Washington

GGülersoy NZ, Balamir M, Bademli R, Türkoĝlu H, Özsoy A, Ünal Y, Erkut G, Eyidoĝan H, Tezer A, Yiĝiter R, Önem B, Arslanlı KY, Çiçek H, Şimşek G, Arslan B, Burnaz M, Şenol M, Taylan A, Özdemir B, Sarı B, Akın U (2003) The earthquake mitigation plan for Istanbul (EMPI)", Boĝaziçi University [BU]; Istanbul Technical University [ITU]; Middle East Technical University [METU]; Yıldız Technical University [YTU], 2003a: Earthquake Master Plan for Istanbul, 262–272. (Istanbul: Metropolitan Municipality of Istanbul, Geotechnical and Earthquake Investigation Department, Planning and Construction Directorate)

İBB (İstanbul Büyükşehir Belediyesi) (2006) 5216 sayılı kanuna göre büyükşehir yetki alanı. [The greater municipality authority area according to law number 5216] http://www.ibb.gov.tr 23 February 2007

İBB (İstanbul Büyükşehir Belediyesi) (2007) Zeytinburnu'ndaki 'deprem dönüşüm' çalışmaları 10 ilçede sürecek [The 'earthquake transformation' projects in Zeytinburnu will continue in 10 districts] http://www.ibb.gov.tr

İBB (İstanbul Büyükşehir Belediyesi) (2009) 1/10,000 Olcekli İstanbul İl Çevre Düzeni Planı [1/100,000 scaled Istanbul Province Environment Regulation Plan]. IBB, Imar ve Sehircilik Daire Baskanligi, Sehir Planlama Mudurlugu, Istanbul. http://www.ibb.gov.tr/TR/Documents/ISTANBUL_CDP_GENEL_BILGI.pdf

Işıngör M (2006) Göçe zorlanan Küçükçekmeceli'nin görüşü alınmadı [The opinion of Küçükçekmece residents that are forced to migrate were not obtained]. Yapı 294:39. Yapı-Endüstri Merkezi, Istanbul

Kabasakal H, Say A, Inelmen K, Yilmaz A (2003) Structural design in the pre- and post-stages of an earthquake: an evaluation of the current administrative structure. BU [Boĝaziçi University]; ITU [Istanbul Technical Univer-sity]; METU [Middle East Technical University]; YTU [Yıldız Technical Univer-sity]: Earthquake Master Plan for Istanbul. Metropolitan Municipality of Istanbul, Geotechnical and Earthquake Investigation Department, Planning and Construction Directorate, Istanbul, pp 215–249

Okten A, Sengezer B, Dinçer I, Batuk G, Koç E, Gül A, Evren Y, Seçkin E, Cekiç TI, Emem O (2003a) Yerleşim çalismaları [Settlement studies], in: Boĝaziçi Üniversitesi (BÜ); İstanbul Teknik Üniversitesi (İTÜ); Orta Doĝu Teknik Üniversitesi (ODTÜ); Yıldız Teknik

Universitesi (YTÜ): İstanbul icin deprem master planı [Earthquake master plan for Istanbul]. İstanbul Büyükşehir Belediyesi, Zemin ve Deprem İnceleme Müdürlüĝü, Planlama ve İmar Dairesi, Istanbul, pp 225–406

Ökten A, Saĝlam F, Şengezer B, Dinçer I, Batuk G, Demir H, Koç E, Gül A, Evren Y, Seçkin E, Çekiç TI, Emem O (2003b) Urban planning, legal issues, administration, finance. Boĝaziçi University [BU]; Istanbul Technical University [ITU]; Middle East Technical University [METU]; Yıldız Technical University [YTU]. Earthquake Master Plan for Istanbul. Metropolitan Municipality of Istanbul, Geotechnical and Earthquake Investigation Department, Planning and Construction Directorate, Istanbul, pp 195–214

Oncüler A (2002) Turkish homeowners' willingness to pay for earthquake mitigation measures. Paper presented at IIASA-DPRI meeting on integrated disaster Risk management, Laxenburg, Austria

Pusch C (2004) Preventable losses: saving lives and property through hazard risk management: a comprehensive risk management framework for Europe and central Asia. Disaster risk management working paper series 9. World Bank, Washington

Saĝlam F, Ökten A, Şengezer B, Dinçer İ, Demir H, Koç E, Gül A, Evren Y, Seçkin E, Çekiç T (2003) Hukuk çalışmaları [Legal studies], in: BÜ [Boĝaziçi Üniversitesi]; İTÜ [İstanbul Teknik Üniversitesi]; ODTÜ [Orta Doĝu Teknik Üniversitesi]; YTÜ [Yıldız Teknik Üniversitesi]: İstanbul icin deprem master planı [Earthquake master plan for Istanbul]. İstanbul Büyükşehir Belediyesi Zemin ve Deprem İnceleme Müdürlüĝü, Planlama ve İmar Dairesi, Istanbul, pp 407–648

Smyth A, Altay G, Deodatis G, Erdik M, Franco G, Gülkan P, Kunreuther H, Luş H, Mete E, Seeber N, Yüzügüllü Ö (2004) Probabilistic benefit-cost analysis for earthquake damage mitigation: evaluating measures for apartment houses in Turkey. Earthq Spectr 20(1):171–203

TCBİB (Turkiye Cumhuriyeti Bayındırlık ve İskan Bakanlıĝı) (2005) Planlama ve imar kanun tasarısı taslaĝı. [Preliminary draft legislation on development and urbanization]. TCBIB, Teknik Araştırma ve Uygulama Genel Müdürlüĝü, Ankara

TMMOB SPO [TMMOB Sehir Plancıları Odası] (2005) İmar kanunu tasarısı taslaĝı ve yeni gelişmeler [Draft legislation on development and new developments]. Statement report. TMMOB SPO, Istanbul

TMMOB Şehir Plancıları Odası (TMMOB ŞPO) (2006a) 1/100.000 ölçekli İstanbul il çevre düzeni planı hakkında TMMOB şehir plancıları odası İstanbul şubesi'nin ön deĝerlendirme raporu [Preliminary evaluation report of the UCTEA chamber of planners about 1/100,000 scaled Istanbul province environment organization plan] Report, July (TMMOB ŞPO İstanbul Şubesi)

TMMOB Şehir Plancıları Odası (TMMOB ŞPO) (2006b) Kentsel dönüşüm deĝil rant amaçlı tasfiye yasası [Not "urban transformation" profit aspired discharge legislation] Press Release (Ankara)

TMMOB Şehir Plancıları Odası (TMMOB ŞPO) (2006c) İstanbul il çevre düzeni planını yargıya taşıdık [We transferred the Istanbul province environment organization plan to the judicatory] E-Bulletin (TMMOB ŞPO İstanbul Şubesi)

TMMOB Şehir Plancıları Odası (TMMOB ŞPO) (2007) Deprem, rant amaçlı kentsel dönüşüm yasasını meşrulaştırmak için kullanılmamalıdır!... [Earthquake should not be used to legalize profit aspired urban transformation legislation] Press Release

Turkish Catastrophe Insurance Pool [TCIP] (unpublished) An overview of the TCIP (Turkish catastrophe insurance pool)

Turkiye Cumhuriyeti İstanbul Büyükşehir Belediyesi (TCİBB) (1995) 1/50.000 ölçekli İstanbul metropoliten alan alt bölge nazım plan raporu [1/50,000 scaled Istanbul metropolitan area sub-region master plan report]. TCİBB, Planlama ve İmar Daire Başkanlıĝı, Şehir Planlama Müdürlüĝü, Istanbul

Türkiye Planlama Okulları Birliĝi (TPOB) (2006) Dönüşüm alanları hakkında kanun tasarısı ile ilgili Türkiye planlama okulları birliĝi tarafindan hazırlanan basın bildirisi [Press release,

prepared by the union of Turkish planning schools, related to draft bill on transformation areas]. Press Release
United Nations Development Programme (UNDP) Turkey (2006) Local administration reform in Turkey. New Horizons UNDP Turkey Monthly Newsletter 3
Uyaroĝlu T (2005) Olası Marmara depremine ilişkin [About potential Marmara earthquake]. Yapı (Istanbul) 288:26–27
WB (World Bank) (2005) Istanbul seismic risk mitigation and emergency preparedness project. http://www.worldbank.org.tr

Chapter 5
Conclusion: Sustainable Urban Development and Good Urban Governance for Disaster Risk Reduction

A major challenge in the application of disaster management policies is the fact that those who benefit and those who pay for it are often not the same people. This commitment of long-term planning and policies lays the basis to the notion of *sustainable development*.

In recent years, the significance of sustainable development for disaster risk reduction has been recognized by international development agencies. In the United Nations' Hyogo Framework for Action, it was stated that "[s]ustainable development, poverty reduction, good governance and disaster risk reduction are mutually supportive objectives" and that "accelerated efforts must be made to build the necessary capacity at the community and national levels to manage and reduce risk" (UN Millenium Project 2005: 7).

Good urban governance is identified to be another important component of disaster risk reduction. As the United Nations' *Local Governments and Disaster Risk Reduction* publication explains "[u]rban risk, city planning and the role of local governments in dealing with risk reduction have been recognized as key factors to build resilient communities and nations since the beginning of the International Strategy for Disaster Reduction" (UN 2010: viii). These factors cannot be more important than today, where more than half of the world's population lives in urban areas.

This book examined the interactions between urban development and urban and risk management in vulnerability and disaster risk reduction. It analyzed the way development of urban areas impacts disaster vulnerability patterns, and how, in return, this vulnerability impacts risk management activities resulting in further inequality, risk and obstacles to sustainable urban development.

The introductory chapter of this book presented an overview of the study, featuring the research questions and outlining research design and methodology. Chapter 2 provided a conceptual foundation for the book. First, it provided the terminology for the study and examined global disaster patterns. Worldwide statistics indicate an increasing number of disasters and disaster impacts (with the exception of mortalities) in recent decades. The global spread of disasters and the

E. A. Gencer, *The Interplay Between Urban Development, Vulnerability, and Risk Management*, Mediterranean Studies 7,
DOI: 10.1007/978-3-642-29470-9_5, © The Author(s) 2013

variety of impacts in different development levels, require the need for diverse policies in disaster risk management.

Chapter 2 continued with the examination of natural disasters in the urban realm. Discussions focused on the global force of urbanization and its relation to climate change. The increasing concentration of population in hazard-prone cities and the embedded conditions of socio-economic and spatial vulnerabilities in urban centers and their peripheries together generate disaster risk in urban areas. With the likely impacts of climate change, such as heat waves or elevation in sea-levels, today, exposure and vulnerability in urban areas deserve a special attention for disaster risk reduction. The study of the urban condition in informal settlements depicted socio-economic, spatial, and institutional factors that result in the vulnerability of the urban poor. This representation showed a strong tie between vulnerability and urban poverty, ranging from physical conditions of living to the disempowerment of residents and their economic poverty. An overview of slum upgrading strategies and contemporary examples of risk reduction programs in informal settlements showed the transformation of the way these settlements have been viewed by local and international communities and its impact on different programs.

Like in many informal settlements, in "formal" urban areas, susceptibility of residents rises predominantly from the physical condition of their living environments. This study indicated that vulnerability is not limited to the poor and that physical planning actions and building regulations and standards alone can help reduce vulnerability in most urban settlements. Discussions in this chapter continued by studying planning tools used for risk reduction by local governments in the United States and by the international development organization and local government collaborations in the international arena. An overview of risk reduction strategies in informal and formal settlements showed that there is no one solution to disaster mitigation in urban areas, and that, different strategies need to be applied to the diverse needs of communities. On the other hand, good urban governance and planning with financial and technical capabilities and with public awareness, empowerment, and participation of urban residents were found to be essential elements in successfully implementing risk reduction strategies in urban areas.

After this theoretical background, Chapter 3 undertook a case study examining the interplay between urban development and vulnerability in the earthquake prone city of Istanbul. After studying the hazard and risk profile of Istanbul, this exploration illustrated pre-earthquake urban planning and development in the city. This investigation focused on understanding the role of urban development in the making of the 1999 earthquake disaster and creating the current socio-economic and spatial vulnerability in the city, presenting the interplay between disaster vulnerability and sustainable urban development.

This analysis revealed that the unsustainable urban development of Istanbul increased the exposure and vulnerability of its residents. The historical overview of urban development and urban planning studies of the Istanbul Metropolitan Area revealed that not paying attention to the rapid increase of migrating populations and their needs, developing narrow-focused urban development plans, and using amnesty laws for political gains resulted in the immense problem of informal settlements throughout the city and facilitated the formation of the socio-economic and physical vulnerability. Additionally, overlooking the results of a geological study and proposing mass-housing and commercial projects in between the two Cekmece lakes in Istanbul resulted in the risk exposure of entire neighborhoods, which were the highest impacted areas in Istanbul by the August 1999 earthquake.

Following up on this disaster, in Chapter 4, the study focused on post-earthquake urban and risk management in Istanbul in order to find out which lessons have been learned and what policy changes have taken place following the 1999 earthquakes. Here the study revealed that both the local and the national government have taken many positive steps after the 1999 earthquake disasters to prepare for and to minimize the impact of an expected earthquake in Istanbul. Among them are the preparations of the comprehensive *Istanbul Earthquake Master Plan* (IEMP), the establishment of regulations that increase building construction quality, the imposition of construction control, and mandatory earthquake insurance. On the other hand, investigations indicated challenges in the implementation of such risk management strategies in the current socio-economic and administrative conditions of the polarized city, in which most residents neither have the means to upgrade, move, or to insure in order to protect themselves.

An empirical research based on interviews in different district municipalities and their risk management activities revealed that, in addition to self-managed coping strategies (such as purchase of insurance), many low-income groups have also been at a disadvantage from a complex urban administration system which has not provided a common public policy, and in which disparities have been witnessed in actions of risk reduction, due to the lack of financial and technical resources. Additionally, an overview of post-urban planning proposals in Istanbul signaled that past practices of short-term planning activities of local administrations with specific political agendas are still exercised by the present-day administrators of the city.

The results stressed once again that the success of urban planning and risk management actions and policies not only lies on providing sustainable solutions to urban dynamics, but also on their execution with good urban governance. As the interplay between urban development, risk management and vulnerability from natural disasters reveals, good urban planning and good urban governance are complementary and essential elements for reducing disaster risks and creating sustainable development in urban areas.

References

United Nations (UN) (2010) Local governments and disaster risk reduction: good practices and lessons learned. UNISDR, Geneva

UN Millenium Project (2005) A home in the city. Task force on improving the lives of slum dwellers. Earthscan, Virginia

Appendix

Istanbul Location Map

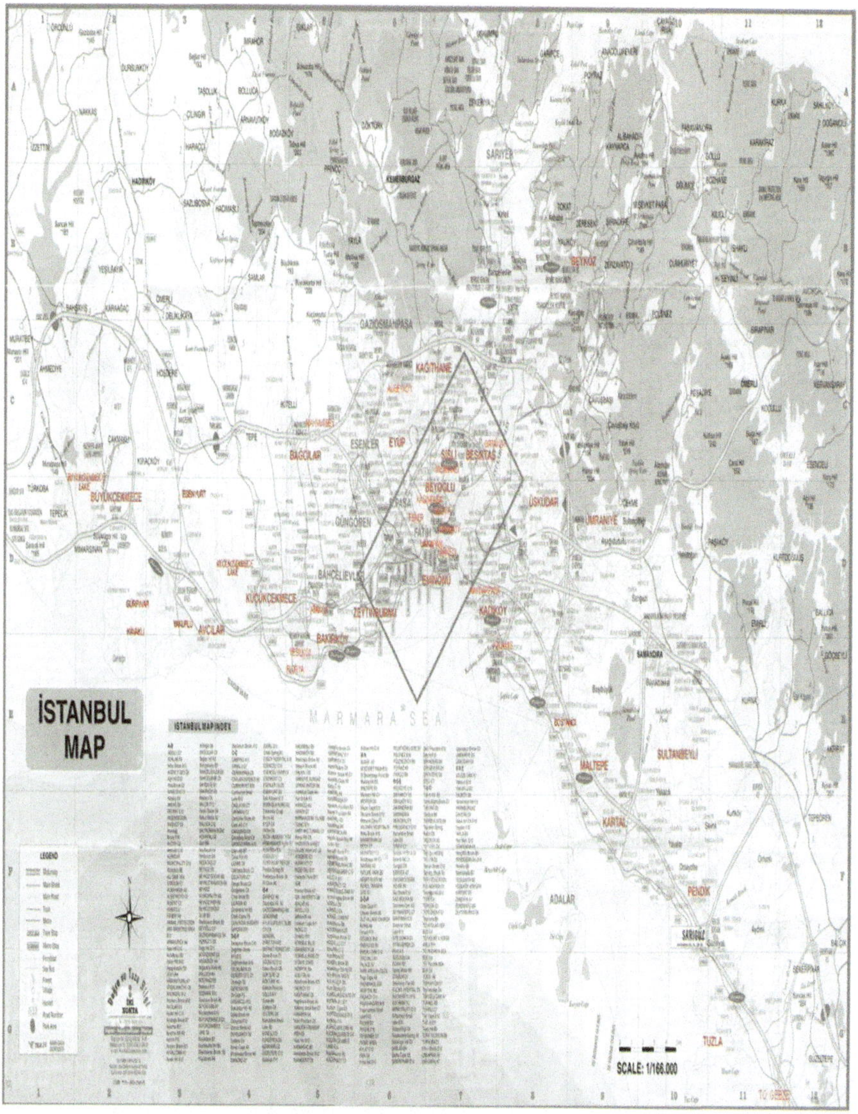

Fig. 1 Istanbul map identifying locations mentioned in Chapters 3 and 4

E. A. Gencer, *The Interplay Between Urban Development, Vulnerability, and Risk Management*, Mediterranean Studies 7,
DOI: 10.1007/978-3-642-29470-9, © The Author(s) 2013

Author Biography

Dr. Ebru Gencer is an urban planner specialized in disaster risk reduction and sustainable urban development (Ph.D, Columbia University, 2007). Ebru Gencer was the lead author of the Natural Hazards and Disaster Risk Management report prepared for the Government of the Dominican Republic by a team of Columbia University scientists and scholars. Previously, she was a researcher at Columbia University Quality Fund Project on Risk Assessment and Mitigation to Metropolitan Areas, taught at Columbia University's Urban Planning Program a graduate studio on Disaster Resilient Planning in Istanbul, and received a ProVention Consortium Applied Research Grant for Disaster Risk Reduction. Dr. Gencer has been a consultant/researcher at Fondazione Eni Enrico Mattei (FEEM). She is a US National Delegate of ISOCARP (International Society of City and Regional Planners), a member of the UNISDR'S Urban Planning Working Group as part of the Making Cities Resilient Campaign, a Euro-Mediterranean and Central America Caribbean think-tank member of the EU (FP7) CATALYST project. Among other scholarly publications, she is recently author of: "The Impact of Globalization on Disaster Risk Trends: A Macro- and Urban-Scale Analysis."

E. A. Gencer, *The Interplay Between Urban Development, Vulnerability, and Risk Management*, Mediterranean Studies 7,
DOI: 10.1007/978-3-642-29470-9, © The Author(s) 2013

The manufacturer's authorised representative in the EU is Springer Nature Customer Service Centre GmbH, Europaplatz 3, 69115 Heidelberg, Germany. If you have any concerns regarding our products, please contact ProductSafety@springernature.com

Printed and bound by CPI Group (UK) Ltd, Croydon, CR0 4YY
30/09/2026
02225575-0010